ELECTRODEIONIZATION

ELECTRODEIONIZATION
Fundamentals, Methods and Applications

P. SENTHIL KUMAR
Centre for Pollution Control and Environmental Engineering, Pondicherry University, Kalapet, Puducherry, India

B. SENTHIL RATHI
Department of Bioengineering, Institute of Biotechnology, Saveetha School of Engineering, Saveetha Institute of Medical and Technical Sciences, Saveetha University, Chennai, Tamil Nadu, India

ELSEVIER

Elsevier
Radarweg 29, PO Box 211, 1000 AE Amsterdam, Netherlands
125 London Wall, London EC2Y 5AS, United Kingdom
50 Hampshire Street, 5th Floor, Cambridge, MA 02139, United States

Copyright © 2024 Elsevier Inc. All rights are reserved, including those for text and data mining, AI training, and similar technologies.

No part of this publication may be reproduced or transmitted in any form or by any means, electronic or mechanical, including photocopying, recording, or any information storage and retrieval system, without permission in writing from the publisher. Details on how to seek permission, further information about the Publisher's permissions policies and our arrangements with organizations such as the Copyright Clearance Center and the Copyright Licensing Agency, can be found at our website: www.elsevier.com/permissions.

This book and the individual contributions contained in it are protected under copyright by the Publisher (other than as may be noted herein).

Notices
Knowledge and best practice in this field are constantly changing. As new research and experience broaden our understanding, changes in research methods, professional practices, or medical treatment may become necessary.

Practitioners and researchers must always rely on their own experience and knowledge in evaluating and using any information, methods, compounds, or experiments described herein. In using such information or methods they should be mindful of their own safety and the safety of others, including parties for whom they have a professional responsibility.

To the fullest extent of the law, neither the Publisher nor the authors, contributors, or editors, assume any liability for any injury and/or damage to persons or property as a matter of products liability, negligence or otherwise, or from any use or operation of any methods, products, instructions, or ideas contained in the material herein.

ISBN: 978-0-443-18983-8

For Information on all Elsevier publications
visit our website at https://www.elsevier.com/books-and-journals

Publisher: Candice Janco
Acquisitions Editor: Anita Koch
Editorial Project Manager: Ashi Jain
Production Project Manager: Rashmi Manoharan
Cover Designer: Mark Rogers

Typeset by MPS Limited, Chennai, India

Contents

About the authors	*ix*
Foreword	*xi*
Preface	*xiii*

1. Introduction   1

1.1	General	1
1.2	Water demand	2
1.3	Various pollutants	4
1.4	Water-energy nexus and water sustainability	13
1.5	Water treatment techniques	14
1.6	Different Electrochemical methods for wastewater treatment along with its advantage and disadvantage	18
1.7	Conclusion	21
	References	21

2. Technology overview of electrodeionization   27

2.1	Introduction	27
2.2	Electrodialysis	28
2.3	Ion exchanger	35
2.4	Electrodeionization	38
2.5	Fundamentals of electrodeionization	44
2.6	Conclusion	48
	References	48

3. Configuration and mechanism of electrodeionization module   53

3.1	Introduction	53
3.2	Electrodeionization mechanism for expulsion and movement of ions	54
3.3	Transport mechanisms created by the ion-exchange resins used in electrodeionization	68
3.4	Principles of adsorption/desorption that affect mass transport in electrodeionization	70
3.5	Conclusion	72
	References	73

v

4. Construction of electrodeionization

79

4.1	Introduction	79
4.2	Overview of electrodeionization	80
4.3	Types of electrodeionization	90
4.4	Chemistry in electrodeionization	95
4.5	Conclusion	97
	References	97

5. Application and comparison of electrodeionization

103

5.1	Introduction	103
5.2	Electrodeionization	104
5.3	Application of electrodeionization	105
5.4	Conventional techniques	109
5.5	Comparison of electrodeionization with other conventional techniques	118
5.6	Comparison of electrodeionization with electrodialysis in terms of cost and energy consumption	119
5.7	Conclusion	123
	References	124

6. Heavy metal ions removal by electrodeionization

129

6.1	Introduction	129
6.2	Heavy metals	130
6.3	Chromium removal by electrodeionization	137
6.4	Arsenic removal by electrodeionization	138
6.5	Cobalt removal by electrodeionization	142
6.6	Nickel removal by electrodeionization	143
6.7	Other metal ions removal by electrodeionization	144
6.8	Conclusion	150
	References	150

7. Electrodeionization in desalination and water softening

155

7.1	Introduction	155
7.2	Desalination	157
7.3	Water softening	168
7.4	Conclusion	177
	References	178

Contents   vii

8. Production of high pure water using electrodeionization   183

8.1 Introduction	183
8.2 Ultrapure water	184
8.3 Application of ultrapure water	186
8.4 Ultrapure water production	189
8.5 Conclusion	199
References	199

9. Advances future scope in electrodeionization   205

9.1 Introduction	205
9.2 Electrodeionization technology	206
9.3 Electrostatic shielding	210
9.4 Electrodeionization reversal	215
9.5 Membrane-free electrodeionization	218
9.6 Resin wafer electrodeionization	224
9.7 Coupling of electrodeionization with other techniques	226
9.8 Artificial intelligence in electrodeionization	228
9.9 Conclusion	229
References	229

10. Economics and environmental aspects of the electrodeionization technique   235

10.1 Introduction	235
10.2 Electrodeionization	236
10.3 Health aspects in electrodeionization	237
10.4 Safety aspects in electrodeionization	239
10.5 Design aspects in electrodeionization	240
10.6 Technoeconomic assessment of electrodeionization	243
10.7 Life cycle analysis of electrodeionization	245
10.8 Conclusion	247
References	247

Index   *253*

About the authors

Dr. P. Senthil Kumar is an Associate Professor in the Centre for Pollution Control and Environmental Engineering, Pondicherry University, Kalapet, India. He has a wide research experience in the fields of water/wastewater. He has 17 years of teaching and research experience and 1 year of industrial experience. He holds 1 Indian Patent on "Shrimp Shell Waste into Biodegradable Grocery Bags," more than 500 International Journal Publications, and more than 100 book chapters and national/international conference publications. He also wrote one book titled "Modern Treatment Strategies for Marine Pollution" published by Elsevier. He has more than 24,500 citations with a Google Scholar h-index of more than 77 along with i10 index of more than 400. He has been listed in Clarivate Highly Cited Researcher 2022 (Cross-Field), Top 2% Scientist reported by Elsevier and Stanford University, United States based on Scopus database for consecutive 3 years from 2020 to 2022, ranked third in India among Best Scientists (Environmental Sciences) for the year 2023 from Research.com, and also received AICTE National Technical Teachers Award (NTTA 2022), Malaviya Memorial (Young Faculty) Award 2019, IEI Young Engineers Award 2017−18, Hiyoshi Think of Ecology Award 2017, CTS Best Teacher Award, and many Young Scientist Awards. He is a recognized Anna University Research Supervisor to guide PhD/MS candidates. Under his guidance, 16 PhD scholars have completed and 10 PhD scholars are currently pursuing in the various fields of environmental and chemical engineering.

Dr. B. Senthil Rathi is an Assistant Professor in the Department of Bioengineering, Institute of Biotechnology, Saveetha School of Engineering, SIMATS, Chennai, 602105, India. She received her MTech degree in chemical engineering and PhD in chemical engineering degrees from Anna University, Chennai, India. She has a wide research experience in the fields of water/wastewater analysis, emerging contaminants removal, adsorption, pollution control treatment methods, and electrochemical treatment methods. He has 10 years of teaching and research experience. She holds 2 patents, more than 20 International Journal Publications, and 4 book chapters. She has more than 900 citations with a Google Scholar h-index of 11.

Foreword

For the existence of life on this planet, water is an incredibly vital supply. The overcrowding of the populace, human activity, rapid industrialization, inadequate exploitation of water resources, and unregulated urbanization have significantly damaged water quality. The demand for water is also increasing due to environmental change, overpopulation, industrialization, and environmental degradation. The public's need to get fresh water for pollutant-free waste disposal via biomass or energy-efficient ways has made industrialized wastewater treatment a top priority. Electrodeionization is a fully advanced ion–exchange method that combines ion exchange, electrodialysis, and elusion procedures for contaminants removal from wastewater. Electrodeionization gained popularity due to the lack of chemicals required for resin regeneration, the production of high purity water, cost-effective methods, and efficient ion removal and recovery.

This book, prepared by Dr. P. Senthil Kumar and Dr. B. Senthil Rathi, will provide readers with fascinating and current developments in the subject of electrodeionization for pollutant removal. Dr. P. Senthil Kumar has over two decades of expertise in pollution remediation using diverse methods such as adsorption, microbial advances, electrodeionization, and others. He is dealing with the environmental issue of disrupting the ecology and its characteristics. Dr. B. Senthil Rathi has greater expertise with adsorption and electrodeionization procedures to remove different contaminants. The writers worked on this book by gathering a large number of articles and technology from various business sectors and published papers. They address the country on various contaminants removal procedures known as electrodeionization, which will assist readers all around the world in learning more about this technology and the many types of pollutants. Readers in many areas can handle these contaminants further by supplying many more new removal properties.

This book will explain issues such as water demand and numerous contaminants that influence the ecosystem. Some of the most recent ways for removing these contaminants are reviewed in detail for simple comprehension. Furthermore, a technical overview of electrodeionization, as well as the setup and operation of the electrodeionization module, is highlighted. With more scientific notions, the construction, use, and comparison of electrodeionization are handled in a broader and neater manner. Overall, the book aims to showcase the most recent breakthroughs and advances in ion-exchange membranes, wastewater zero discharge based on ion-exchange membranes, membrane free, and resin wafer electrodeionization cells.

Preface

Purpose

The information in the book is suitable for beginners who want to learn about the procedures used to remove various contaminants via electrodeionization. This book is excellent for self-study as well as a reference study for those working in this industry. Many of the techniques for treatment are explored in depth with relevant topics. Current approaches used across the world to remove various contaminants employing electrodeionization and their progress are addressed and illustrated in an easy-to-understand way.

Background

It is not necessary to be familiar with electrodeionization. This book may be used with just the concepts of various contaminants. This book contains detailed information on the removal of various contaminants using various procedures, notably electrodeionization. It may be regarded as a mini-review highlighting the significance of the electrodeionization technology for eliminating pollutants due to its numerous advantages.

Organization

Before starting every chapter, an abstract covering the topics that will be explored in detail inside the chapter is included. The mentioned contents emphasize the subdivision that is covered throughout the chapters. Each of the concepts is emphasized with references to the chosen and discussed point. Each chapter concludes with conclusions that summarize what has been addressed. Users are already familiar with the technology being addressed, allowing them to be improved further.

Concept

The major subject covered in this book is the treatment technologies that are utilized across the world to safely remove various sorts of contaminants. The book also depicts some of the consequences of various contaminants. It demonstrates the principle, deployment, and comparison of electrodeionization with various alternatives. Chapters 6–9 showcase some of the most significant applications and recent advances in electrodeionization.

CHAPTER ONE

Introduction

1.1 General

A crucial worldwide concern, the supply of pure water available for human use will be made worse by the consequences of climate change (Russell and Fielding, 2010). Portable water is becoming a limited commodity in various global areas due to the population explosion, water shortages, and global warming-related protracted dry spells and floods. The water business has to produce reliable, affordable, and efficient technologies and ways for supplying enough freshwater. Due to the rising water demand, strict health regulations, and growing pollutants, traditional techniques of treating water and sewage water are nevertheless insufficient to give sufficient access to safe drinking water (Amin et al., 2014). As a result, the cost-effective and efficient elimination of pollutants from water has received attention. Heavy metal ions, radioactive materials, hydrocarbons, fertilizers, medicines, dyes, and other contaminants are among the most prevalent ones found in wastewater. According to their chemical constitution, these contaminants might be identified as either organic or inorganic (Gao et al., 2019).

Electricity and water are inherently related. Fundamentally, electricity is required for both water purification and transportation, while water is required for energy production. Due in large part to the assumption that neither water nor electricity posed a danger to either security of supply, there has traditionally been little incentive to comprehend the nature of these connections. This assumption is currently being contested. The connections among water and power are coming into sharp light in previously unheard of ways due to the changes in the industry, increasing preference, and much more lately, climate variability. Limited knowledge of the nature of the connections and a perceived lack of policy instruments to adequately assess them are impeding society's ability to cope with the difficulties and ambiguities brought on by the connections among electricity and water (Hamiche et al., 2016). Furthermore to ensure the long-term viability of the potable water for rising water demand, taking the water-energy nexus into account when optimizing water systems also reduces water-related environmental and energy problems (Vakilifard et al., 2018).

Researchers are seeking for low-cost and appropriate solutions to clean and recycle wastewater because it is a major problem. Groundwater reduction, treatment of wastewater, and reuse are the three major reasons for which water treatment techniques are utilized (Gupta et al., 2012). In order to eliminate pollutants from wastewater, physical, chemical, and biological approaches are typically employed in wastewater treatment, modern methods for removing contaminants, such as photocatalysis, chemical precipitation, flotation, electrochemical treatment, bioremediation, membrane separation techniques, adsorption, ion exchange, coagulation, flocculation, and so on. While selecting the most effective technique for pollutants removal, the associated costs, the characteristics of the effluent, and the method suitability are typically utmost crucial factors (Bhargava, 2016; Saleh et al., 2022).

Electrochemical treatments provide a number of benefits over other techniques for treating wastewater. They are reliable, simple to use, and adaptable to changing wastewater flow conditions. Additionally, a significant range of contaminants both organic and inorganic may be eliminated (Muddemann et al., 2019). Today's electrochemical technologies have advanced to the point where they are not only cost-competitive with conventional systems as well as more effective and portable. In some cases, treating wastewaters containing resistant contaminants may need electrochemical techniques (Chen, 2004). This chapter investigates the demand for water as well as the necessity for wastewater reuse. Water-energy nexus, water sustainability, and various treatment procedures have all been investigated as potential contaminants in wastewater. Additionally, several electrochemical approaches, together with their benefits and drawbacks, are addressed toward the chapter's conclusion.

1.2 Water demand

Among the biggest problem confronting the world right now is the pollution problem, according to global warming estimates, it will only get worsen in the next years. For areas with water crisis, finding a solution is crucial because accessibility to and access to water are the major restraints on crop yields (Mancosu et al., 2015). Over than a third of those living in developing nations or even more than a 1/4 of the worldwide people now reside in areas that will have severe water shortages by the turn of the century (Seckler et al., 1999). Although it is a sustainable option, freshwater has a limited supply. There are now countless indications from all across the globe that human water usage is excessive. The most evident signs of water stress include depletion of natural resources, low or absent river flows, and rising pollution levels. In several places, increasing water extraction for human consumption endangers the wellbeing of crucial aquatic life (Postel, 2000). Household potable water demand is affected by economic, social, and environmental aspects, all of which are predicted to alter significantly in the near future. Particularly, sewage costs may rise due to environmental regulations to limit dangerous chemicals, water rates may alter if water marketplaces are unregulated, or they may increase in response to growing shortages or maintenance and rebuilding demands. Higher family incomes brought on by economic expansion may increase demand for extra water services. In addition to changing temperatures and rainfall variability throughout time, climate change may also have an impact on household water usage, particularly if gardening irrigation requirements vary (Schleich and Hillenbrand, 2009). To maintain cities and ensure societal water security, a public water system must be managed properly. Urban freshwater estimate has historically been a challenging issue for legislators and utility company managers (Zubaidi et al., 2020).

1.2.1 Water demand reduction strategy

Changing climate has emerged as a daily reality for mankind as a consequence of the expansion of many businesses and practices that raise emission of greenhouse gases. Global food security, public health, and the integrity and accessibility of water resources are all significantly harmed by climate change. Besides other factors, the rise in population, urbanization, usage per person, water contamination, and climate change all have an effect on readily available aquifers. Water shortage is a major indication of illness and a problem of poverty that primarily affects residents of rural regions where there are significant population concentrations. Large amounts of wastewater are produced everyday in agriculture, industry, and residences. The amount of wastewater used in private households accounting for 50%—80% of all water usage, and, as stated previously, the worldwide wastewater discharge was calculated at 400 billion cubic meters per year, contaminating around 5500 billion cubic meters of water annually. Typically, wastewater is composed of 99% water and 1% dissolved, suspended, and colloidal particles (Ungureanu et al., 2020).

Since water is a precious resource in many regions of the globe, it is important to develop creative methods to clean wastewater so that it may be recycled. The most popular and recommended approach is to treat wastewater for reuse (Jaiyeola and Bwapwa, 2016). Reusing effluent has been acknowledged as a promising option to deal with the issue of water shortage on a global scale. The execution of wastewater reuse projects would be certain to take into consideration all the different types of impacted customers, as well as the external benefits and costs obtained from the recycling choice, using Combined Water Resources Management concepts (Garcia and Pargament, 2015; Duong and Saphores, 2015). Reusing gray water has been shown to be a dependable alternative water supply that may be an important part of integrated water resource management and offers a practical remedy for dealing with water constraint situations. Reclaimed water was expected to make a considerable contribution to reducing water shortages in the islands with the highest potential for reuse, suggesting that it would be a useful strategy for reducing water scarcity (Stathatou et al., 2015). Wastewater is one of the most reliable sources of freshwater. Commercial, agricultural, and household activities all rise in tandem with population growth in order to meet the human's requirements. Large amounts of wastewater are produced by these operations, yet water may still be recovered and used for a variety of other things. The effectiveness of typical sewage treatment techniques in processing wastewaters for discharge over period is restricted. Yet, to produce treated wastewater useable for commercial, farming, and home applications, advances in wastewater treatment technologies are required (Obotey Ezugbe and Rathilal, 2020).

1.3 Various pollutants

The most significant and necessary commodity on planet is water. Nevertheless, industrialization, home use, and farming activities are all contributing to the ongoing decline in water quality of our water resources. Numerous organic pollutants, including pharmaceuticals, pesticides, hydrocarbons, fertilizers, plasticizers, greases, phenols, detergents, biphenyls, oils, etc., in addition inorganic pollutants, such as fluoride, sulfate, phosphate, heavy metals, and nitrate, addition biological pollutants, including bacterium, microbes, microalgae, planktons, pathogens, fungus, and others, have all been identified as contaminants in water. For the removal of pollutants, several techniques have been created and utilized (Wang and Zhuang, 2017).

1.3.1 Dyes

Dyeing has a long tradition and plays a significant role in our everyday lives. The dye business initially relied on organic plant and insect sources before quickly switching to synthetic production techniques. Sadly, a number of synthetic dyes, particularly azo dyes, has shown to be toxic and mutagenic and therefore are currently outlawed globally. Nonetheless, synthetic dyes are mostly made and utilized today because of their low cost as well as other desirable characteristics. Azo dye removal and treatment from wastewater is a highly unique task. As dyestuffs are prone to the standard conventional biological treatment and physiological treatment processes are hardly cost-effective not ecologically safe, innovative treatment solutions must be researched (Bafana et al., 2011).

One of the major businesses that annually produces a significant quantity of industrial waste that is the primary cause of water pollution that is toxic to aquatic life and carcinogenic to humans is the textile industry (Hassaan et al., 2017). A long-term environmentally friendly and effective dye wastewater treatment approach must be developed to address this issue. Dye wastewater needs to be treated before discharge in order to decrease its negative impact on the ecosystem as well as other living creatures. Yet, because there is a lack of information on efficient dye extraction methods, it is difficult to choose a specific remedy for the existing dye discharge issue (Katheresan et al., 2018). Nearly, all physical, chemical, and biological techniques, such as flocculation, coagulation, adsorption, membrane separations, oxidation, ion-exchange, electrochemical process, advanced oxidation process, biodegradation, photocatalysis, and others, are used in the treatment of dyeing wastewater (Zhou et al., 2019). Table 1.1 gives the removal percentage of dyes using adsorption along with its process condition.

Table 1.1 Removal percentage of dyes using adsorption.

Adsorbent	Contaminant	Initial concentration	pH	Adsorbent dosage	Isotherm	Adsorbed per unit mass of adsorbent (mg/g)	References
Aluminum-based pillared clay	Methylene Blue	30 mg/dm^3	7	0.1 g/10 cm^3	Freundlich Isotherm	2.2	Gil et al. (2011)
Aluminum-based pillared clay	Orange II	30 mg/dm^3	7	0.1 g/10 cm^3	Sips Isotherm	9.8	Gil et al. (2011)
Bark	Methylene Blue	100 mg/dm^3	7	0.1 g/0.050 dm^3	Langmuir isotherm	914	McKay et al. (1999)
Bark	Safranine	100 mg/dm^3	7	0.1 g/0.050 dm^3	Langmuir isotherm	875	McKay et al. (1999)
Cellulose-based bioadsorbent	Anionic ye acid blue 93	200 ppm	9	0.4 g/200 mL	Freundlich Isotherm	1372	Liu et al. (2015)
Cellulose-based bioadsorbent	Cationic dye methylene blue	200 ppm	9	0.4 g/200 mL	Freundlich Isotherm	1372	Liu et al. (2015)
Coal	Methylene Blue	100 mg/dm^3	7	0.1 g/0.050 dm^3	Langmuir isotherm	250	McKay et al. (1999)
Coal	Safranine	100 mg/dm^3	7	0.1 g/0.050 dm^3	Langmuir isotherm	120	McKay et al. (1999)
Cotton	Methylene Blue	100 mg/dm^3	7	0.1 g/0.050 dm^3	Langmuir isotherm	277	McKay et al. (1999)
Cotton	Safranine	100 mg/dm^3	7	0.1 g/0.050 dm^3	Langmuir isotherm	838	McKay et al. (1999)
Hair	Methylene Blue	100 mg/dm^3	7	0.1 g/0.050 dm^3	Langmuir isotherm	158	McKay et al. (1999)
Hair	Safranine	100 mg/dm^3	7	0.1 g/0.050 dm^3	Langmuir isotherm	190	McKay et al. (1999)
Rice husk	Methylene Blue	100 mg/dm^3	7	0.1 g/0.050 dm^3	Langmuir isotherm	312	McKay et al. (1999)
Rice husk	Safranine	100 mg/dm^3	7	0.1 g/0.050 dm^3	Langmuir isotherm	1119	McKay et al. (1999)
Zirconium-based pillared clay	Methylene Blue	30 mg/dm^3	7	0.1 g/10 cm^3	Freundlich Isotherm	7	Gil et al. (2011)
Zirconium-based pillared clay	Orange II	30 mg/dm^3	7	0.1 g/10 cm^3	Sips Isotherm	39	Gil et al. (2011)

1.3.2 Heavy metals

Although the geological systems and biological balance of heavy metals in the earth's mantle are not normally altered, reckless human action has drastically altered these phenomena. As a result, metals build up in parts of the plant that also include chemical constituents, which results in a particular inhibitory effect. After being exposure to heavy metals such as Cu, Cd, Ni, Zn and Pb over a long duration, people may experience unfavorable health effects (Singh et al., 2011). Metals may interact with a broad range of organic molecules. Both sensitive and insensitive locations were identified to jointly host heavy metal binding (Passow et al., 1961). The main way that people are subjected to toxic metals is via water that has been polluted. The effects of drinking water that have been contaminated with toxic metals including Pb, As, Cd, Ni, and Hg have gradually come to the attention of the relevant authorities and staff. It is well accepted that employees are subjected to toxic metals when these metals are used in a variety of production processes and/or a variety of products, such as dyes and composites. Access to drinking water polluted with heavy metals has been linked to a number of harmful consequences on human metabolism that have been documented from all over the world (Fu and Xi, 2020).

With different degrees of success, substantial research has been done over many decades to find a simple, effective, and affordable way to remove heavy metals (Shrestha et al., 2021). The most widely used traditional procedures for the expulsion of these heavy metals are adsorption, flocculation, chemical precipitation, ion exchange, ion floatation, coagulation, and electrochemical approaches (Fu and Wang, 2011). Table 1.2 gives the adsorption's effectiveness in removing heavy metals, which is dependent on several factors, including pH, starting concentration, isotherm models, and others. These techniques, however, have several severe drawbacks, including high sludge generation that necessitates additional treatment, limited highest removal, and large energy needs. Newer, more effective, more affordable, and inventive technologies are now being studied. Electrodialysis, hydrogels, photocatalysis, membrane separation techniques, and the use of novel adsorbent materials have recently emerged for improved adsorption (Azimi et al., 2017).

1.3.3 Emerging contaminants

The existence of emerging or newly identified contaminants in our water sources remains to cause concern regarding the health and welfare of the general people. The standard water treatment facilities that are currently in place were

Table 1.2 Removal percentage of heavy metals using adsorption.

Adsorbent	Contaminant	Initial concentration (ppm)	pH	Adsorbent dosage	Isotherm	% Removal/ adsorbed per unit mass of adsorbent (mg/g)	References
Walnut Saw dust	Cadmium	100	5	2 g/100 mL	Langmuir and Freundlich isotherms	4.39 mg/g	Bulut (2007)
Coirpith	Cadmium	700	4	300 mg/50 mL	—	100%	Kadirvelu et al. (2001)
Pleurotus ostreatus	Chromium	93.50	4.5—5.0	0.2 g/100 mL	Langmuir and Freundlich isotherms	4.45 mg/g	Javaid et al. (2011)
Pleurotus ostreatus	Copper	23.56	4.5—5.0	0.2 g/100 mL	Langmuir and Freundlich isotherms	2.73 mg/g	Javaid et al. (2011)
Coirpith	Copper	126	5	450 mg/50 mL	-	73%	Kadirvelu et al. (2001)
Tire-derived Activated Carbon	Copper	500	1.5	0.2 g/50 mL	Langmuir isotherms	185.2 mg/g	Shahrokhi-Shahraki et al. (2021)
Commercial activated carbon	Copper	500	1.5	2.5 g/50 mL	Freundlich isotherms	15.0 mg/g	Shahrokhi-Shahraki et al. (2021)
Walnut Saw dust	Lead	100	5	2 g/100 mL	Langmuir and Freundlich isotherms	6.54 mg/g	Bulut (2007)
Coirpith	Lead	709	4	300 mg/50 mL	—	100%	Kadirvelu et al. (2001)
Tire-derived Activated Carbon	Lead	500	1.5	0.2 g/50 mL	Langmuir isotherms	322.5 mg/g	Shahrokhi-Shahraki et al. (2021)

Adsorbent	Metal				Isotherm	Capacity	Reference
Commercial activated carbon	Lead	500	1.5	2.5 g/50 mL	Freundlich isotherms	42.5 mg/g	Shahrokhi-Shahraki et al. (2021)
Coirpith	Mercury	100	3.5	125 mg/50 mL	—	100%	Kadirvelu et al. (2001)
Walnut Saw dust	Nickel	100	5	2 g/100 mL	Langmuir and Freundlich isotherms	2.40 mg/g	Bulut (2007)
Pleurotus ostreatus	Nickel	54.83	4.5−5.0	0.2 g/100 mL	Langmuir and Freundlich isotherms	8.45 mg/g	Javaid et al. (2011)
Coirpith	Nickel	996	3.5	250 mg/50 mL	—	92%	Kadirvelu et al. (2001)
Pleurotus ostreatus	Zinc	42.87	2.5	0.2 g/100 mL	Langmuir and Freundlich isotherms	0.88 mg/g	Javaid et al. (2011)
Tire-derived Activated Carbon	Zinc	500	1.5	0.2 g/50 mL	Langmuir isotherms	71.9 mg/g	Shahrokhi-Shahraki et al. (2021)
Commercial activated carbon	Zinc	500	1.5	2.5 g/50 mL	Freundlich isotherms	14.0 mg/g	Shahrokhi-Shahraki et al. (2021)

not intended to handle these unknown chemicals. Pharmaceuticals, personal care items, surfactants, different industrial additives, and several substances that are allegedly endocrine disruptors are included in the category of endocrine disrupting chemicals. These are currently a risk to our system of water supply. Due to the lack of strict regulations that are unique to these pollutants, the present wastewater treatment system is ineffective in eliminating these various classes of developing contaminants. Unwanted substances are being discharged into the aquatic system, either purposefully or inadvertently, and they have an impact on every living thing (Bolong et al., 2009). The emergence of extremely sensitive and potent analytical instruments that quickly evolved, allowing detection and minimal measurement, has helped to advance our knowledge of emerging pollutants by revealing previously undiscovered pollutants in intricate environmental media. The cornerstones of environmental studies and high-performance chromatographic separations combined with high-resolution mass spectrometers are becoming more prevalent in ecologic laboratories. These methods help us to learn more about the existence of emerging pollutants in the environment, how they change and end up there, and the intricate ecological effects they have on exposed biological systems (Noguera-Oviedo and Aga, 2016).

As a way to eliminate developing contaminants from all adsorption stands out, several techniques have been tried, including membrane technology, sophisticated oxidation processes, and biological ways. Adsorption is a good strategy for eradicating new contaminants because it has a clear operating setup, fairly low cost of installation, and great effectiveness (Rathi et al., 2021). Table 1.3 depicts the removal percentage of emerging pollutants using adsorption along with its pH, initial concentration, isotherms, and other conditions. Pharmaceuticals, beta-blockers, pesticides, and endocrine disrupting chemicals were among the range of emerging contaminants that were successfully removed by the membrane bioreactor and membrane filters working together in a hybrid model. Certain coupled technologies composed of constructed wetlands and sludge ponds demonstrated significant promise for the biosorptive removal of drugs and specific beta-blockers. Composite systems that integrate the activated sludge technique with physical procedures such as ultrafiltration, reverse osmosis, and gamma ray are the most successful and effective ways to remove trace organic pollutants. The hybrid version of MBR coupled with UV degradation, activated charcoal and ultrasonic, and ozonation accompanied by ultrasounds completely eradicated certain emerging pollutants and several medications (Dhangar and Kumar, 2020). Combining electrochemical oxidation with membrane separation has been shown to

Table 1.3 Removal percentage of emerging contaminants using adsorption.

Adsorbent	Contaminant	Initial concentration (mg/dm^3)	pH	Adsorbent dosage	Isotherm	% Removal/ adsorbed per unit mass of adsorbent (mg/g)	References
Activated carbon	Clofibric acid	15	3	25 mg/L	Sips isotherm	5.2	Taoufik et al. (2019)
Activated carbon	Gallic acid	15	3	25 mg/L	Sips isotherm	53	Taoufik et al. (2019)
Activated carbon	Salicylic acid	15	3	25 mg/L	Sips isotherm	163	Taoufik et al. (2019)
Activated carbon coated 1% titanium dioxide	Clofibric acid	15	3	25 mg/L	Sips isotherm	27	Taoufik et al. (2019)
Activated carbon coated 1% titanium dioxide	Gallic acid	15	3	25 mg/L	Sips isotherm	14	Taoufik et al. (2019)
Activated carbon coated 1% titanium dioxide	Salicylic acid	15	3	25 mg/L	Sips isotherm	659	Taoufik et al. (2019)
Activated carbon coated 10% titanium dioxide	Clofibric acid	15	3	25 mg/L	Sips isotherm	56	Taoufik et al. (2019)
Activated carbon coated 10% titanium dioxide	Gallic acid	15	3	25 mg/L	Sips isotherm	144	Taoufik et al. (2019)
Activated carbon coated 10% titanium dioxide	Salicylic acid	15	3	25 mg/L	Sips isotherm	990	Taoufik et al. (2019)
Activated carbon coated 5% titanium dioxide	Clofibric acid	15	3	25 mg/L	Sips isotherm	15	Taoufik et al. (2019)

(Continued)

Table 1.3 (Continued)

Adsorbent	Contaminant	Initial concentration (mg/dm^3)	pH	Adsorbent dosage	Isotherm	% Removal/ adsorbed per unit mass of adsorbent (mg/g)	References
Activated carbon coated 5% titanium dioxide	Gallic acid	15	3	25 mg/L	Sips isotherm	20	Taoufik et al. (2019)
Activated carbon coated 5% titanium dioxide	Salicylic acid	15	3	25 mg/L	Sips isotherm	669	Taoufik et al. (2019)
Commercial activated carbon (CAC)	Caffeine	15	6	50 mg/10 cm^3	Toth Isotherm	112	Gil et al. (2018)
CAC	Clofibric acid	15	6	50 mg/10 cm^3	Toth Isotherm	—	Gil et al. (2018)
CAC	Diclofenac	15	6	50 mg/10 cm^3	Toth Isotherm	2557	Gil et al. (2018)
CAC	Galli acid	15	6	50 mg/10 cm^3	Toth Isotherm	—	Gil et al. (2018)
CAC	Ibuprofen	15	6	50 mg/10 cm^3	Toth Isotherm	247	Gil et al. (2018)
CAC	Salicylic acid	15	6	50 mg/10 cm^3	Toth Isotherm	422	Gil et al. (2018)

CAC, commercial activated carbon.

be effective in reducing the main limitations of both processes, such as membrane technique blockage and mass transfer limitations in the electrolytic oxidation process (Shahid et al., 2021).

1.3.4 Persistent organic pollutants

Owing to their bioaccumulative and long-lasting characteristics, persistent organic pollutants are known as the real killers. All living things, including humans, mammals, and plants, have them. These are in charge of many deadly illnesses and environmental issues. The various illnesses brought on by persistent organic pollutants include diabetic, overweight, endocrine disruption, cancer, cardiac, fertility, and environmental issues. Alarm over persistent organic pollutants contamination and risks is high among academics, policymakers, and NGOs (Alharbi et al., 2018). There is growing worry over the possible dangers of exposure in humans to POPs because of their widespread occurrence in the atmosphere and lipophilic characteristics (Qing Li et al., 2006). Persistent organic pollutant measurements and data on their dispersion and ecological factors are both crucially dependent on analytical chemistry. The creation of these contaminants' analytical procedures that are quicker, safer, more dependable, and more accurate has throughout the 20 years prior it has drawn considerable interest. Analytical techniques have advanced significantly since the Stockholm Convention on persistent organic pollutants was enacted 12 years ago (Xu et al., 2013).

1.4 Water-energy nexus and water sustainability

Water supply reliability in context of growing demand is ensured by taking the water-energy connection into account while optimizing water systems, which also reduces water-related energy and ecological problems (Vakilifard et al., 2018). The extensive examination of connections between water and energy termed as the water-energy nexus has garnered a lot of fascination in the recent days owing to the rising anxiety for the security of both resources throughout the world. Water-energy nexus assessment is thought to help with more informed management for water and energy as well as a better grasp of potential solutions for both legislation and innovation. This should help policymakers and funding agencies to conserve water and energy and promote sustainability (Dai et al., 2018).

Techniques for desalination and effluent treatment that use less energy are essential for increasing freshwater resources without unduly taxing the world's finite energy sources. Using limited or waste energy to desalinate saltwater from high-salinity fluids, membrane distillation has the potential to boost sustained freshwater generation, an important component of the water-energy nexus. Despite advancements in membrane science and the creation of novel process setup, the viability of membrane distillation as an energy-efficient desalination method is still under question (Deshmukh et al., 2018). Water-free energy production, carbon dioxide sequestration, low-energy separations, and improved fluids are top research goals in the water-energy nexus (Urban, 2017).

Due to the global lack of supplies, it will put tremendous strain on the current water and energy systems. According to reports, the largest global concern is the environmental problem brought on by irresponsible water and energy use. Here, preserving water and energy rank among the top concerns for global sustainable development. The interdependence of the water and energy systems is much more significant. On the one hand, the principal means of generating, transporting, and utilizing energy in the energy system is water. Because the interdependence of water and energy resources, saving energy can ease strain on water resources while simultaneously reducing energy usage (Ding et al., 2020).

Environmental sustainability has been severely strained by higher efficiency of energy, saving water, carbon-free emissions, waste disposal, and preserving food issues. For many extended water-energy nexuses, input—output analysis and life cycle assessment must be effectively expanded or combined. It is necessary for analyzing inventory flow across various places and businesses, as well as for figuring out the associated environmental and economic implications. The nexus may prevent becoming a buzzword group by adopting standardized understandings such generating life cycle assessments and even more quantified footprints, which would support sustainable development objectives (Wang et al., 2021).

1.5 Water treatment techniques

Water pollution and its management have become a growing problem around the world. Extreme measures have been taken lately to solve the problems related to wastewater treatment. There are several recognized methods for treating effluent, including physical methods such as adsorption and membrane filtration, chemical methods such as Fenton oxidation and electrochemical oxidation, and several biologically-derived (Rashid et al., 2021).

1.5.1 Adsorption

Adsorption is regarded as among the most well-liked and often applied heavy metal removal strategies among all approaches. This approach has the benefits of being straightforward, simple to use, inexpensive, and neutral to hazardous chemicals and toxic contaminants. Finding and creating highly trustworthy adsorbents has always been a key endeavor if they are to be widely used in industries and made commercially available. Chitosan, silica, activated carbon, zeolite, carbon nanotubes, and cellulose are a few well-known adsorbents with a focus on the eradication of toxic metals. The majority of these sorbent materials are porous inorganic or organic compounds with a binding specificity for adsorbing toxic metals (Ahmad et al., 2020).

Modern adsorbents such as metal oxides and related composite materials have lately attracted a lot of interest. Due to their abundance of surface-active sites, easily controlled surface characteristics, straightforward synthesis and functionalization, high attainable surface area, commercial viability, and high recyclability, metal oxide-based nanomaterials are likely materials for the rapid and efficient expulsion of a wide range of heavy metal and metalloid ions. The most recent developments in adsorption are shown in Fig. 1.1. The technical challenges and forward-looking guidelines are highlighted in the interest of enhancing the adsorption effectiveness and renewability of metal oxide-based nanoparticles, like their use at an enormous scale to illustrate their economic viability for useful applications, a motive for water security (Gupta et al., 2021).

1.5.2 Membrane technology

The membrane separation technique, which has been viewed as a viable method for removing contaminants, contains some of its best qualities, including good efficiency, simple operation, and minimal area requirements (Xiang et al., 2022). Researchers working on a range of maritime uses, such as the sewage treatment, water purification, water sterilization, toxic metals, harmful and nontoxic chemical compounds, among others, have shown a great deal of curiosity in membranes. Membranes serve as the main element of membrane-based methods for detachment (Castro-Muñoz et al., 2021). However, disposal and clogging, correspondingly, are the key problems that often involve adsorption and membrane separation.

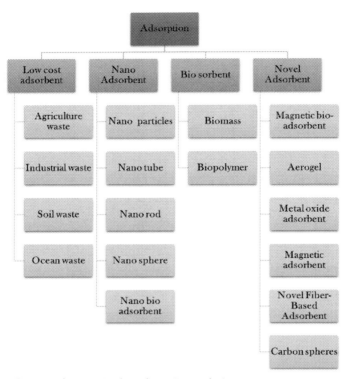

Figure 1.1 Recent advances in the adsorption technique.

The adsorption method is a viable alternative because it has a number of advantages over membrane techniques, including low startup and operating costs, ease of use, and the ability to remove dangerous substances using a range of solid media (Albatrni et al., 2021). Fig. 1.2 represents the recent advances in the membrane technology. In conclusion, membrane technology, particularly NF, has recently made tremendous progress toward the eradication of artificial colors from wastewater. To make membrane technology more effective in the industrial sector, more work, study, and research are needed. Research on using RO membranes to remove color from wastewater is yet inadequate (Moradihamedani, 2022).

1.5.3 Biological methods

In the methods of biological treatment, the metabolic capacities of bacterium, protozoa, fungi, microalgae, or vegetation are exploited to

Introduction

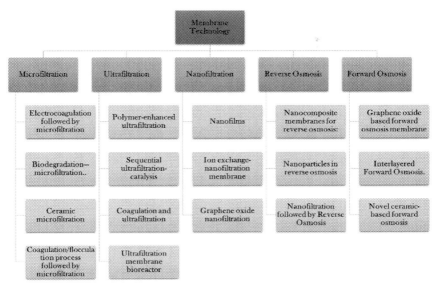

Figure 1.2 Recent advances in the membrane technology.

oxidation or reduce substances that are both organic and inorganic. By maintaining ideal circumstances, such as oxygen content, pH, availability to nutrients, and so on, the mechanisms that take place in nature are accelerated. They are typically carried out in bioreactors with suitable aeration and agitation apparatus; however, occasionally wetlands are employed. In terms of operational expenses, it is believed that the biodegradation processes are economically advantageous. However, based on the equipment utilized, investment costs might be significant. Bioprocesses have a number of constraints, such as the fact that they can only alter biodegradable substances. Moreover, toxic substances may impair biological processes and potentially limit the utilization of plants or bacteria (Paździor et al., 2019). Bacteria and other organisms utilize and change the organic pollutants that must be eliminated as a source of sustenance. Therefore, waste is disposed of in a nontoxic and hygienic way using the water/wastewater treatment procedure. When treating water and wastewater, biological processes come in two basic categories. The microbes in charge of the treatment process are kept in fluid medium in the first of these procedures, which is also referred to as suspended growth. However, in attached growth (biofilm) methods, the microbes in charge of converting organic matter or nutrition are bonded to an inactive carrier materials such as stone, grit, or ash, or to a variety of different polymers and other synthetic fabrics (Yildiz, 2012).

1.5.4 Coagulation and flocculation

Coagulants that were often utilized were ferric chloride, aluminum sulfate, ferrous sulfate, and ferric chloro-sulfate. Yet, a challenge in treating wastewater has emerged due to a possible link between Alzheimer's disease and conventional aluminum-based coagulants. The use of the more ecologically friendly biopolymer chitosan in treatment has so become more important. In addition, deacetylating chitin yields the natural organic polyelectrolyte chitosan, which has a large molecular weight and maximal charge density (Prakash et al., 2014). Modern wastewater treatment methods that are both popular and efficient mainly rely on flocculation and coagulation. Their goal is to improve particle species segregation in downstream processes such as filtration and sedimentation. In order to create bigger size particulates that can be removed more effectively, colloidal particles and other tiny separated substances are gathered together and aggregated (Shammas, 2005). Hybrid materials are those created by combining useful components with the basic material to increase the accumulating power. They are employed in the coagulation/flocculation of effluent. It makes sense to add multifunctional chemical groups or elements to the base chemical to increase the aggregation potential. Hybrid materials therefore exhibit greater efficiency compared to those made composed of separate components because of the synergistic impact of hybrid components in one substance. Instead of the conventional coagulation-flocculation system's two procedures, the full treated water can be processed with the furthermore of a single compound and in a single tank. As a result, hybrid substances, which combine multiple core elements into one medication, may be a convenient and effective option available for the process of water treatment plants (Lee et al., 2012).

1.6 Different Electrochemical methods for wastewater treatment along with its advantage and disadvantage

When it pertains to sewage treatment, electrochemical procedures provide a wide range of benefits over conventional methods. These methods are reliable, easy to use, and flexible enough to react to shifting effluent streams. Additionally, a variety of pollutants can be removed. The benefits of electrochemical technique for treating wastewater are its maximum removal efficacy, green power generation, minimal environmental effect, ease of operation, and compact size. Electrochemical processes including electroflotation, electrocoagulation, electrodialysis, electrodeionization, electroreduction, electrooxidation, and electrodisinfection have been significantly improved (Yadav and Kamsonlian, 2022; Mook et al., 2014).

1.6.1 Electrocoagulation

With the use of electrocoagulation technologies, pollutants may be treated and flocculated without the need for additional coagulants. Because direct current is used to move the microscopic particles, coagulation happens as the current is applied, making it possible to remove them. Electrocoagulation may also lower the amount of residue needed to produce garbage. Anodes and cathodes, two sets of metal plates used in electrocoagulation, are placed in pairs of two. Using the electrochemical rules, the cathode is oxidized while the water is reduced to enhance treated wastewater. When the cathode electrode makes contact with the effluent, metal is discharged into the apparatus. When this occurs, hydroxide complexes are formed in order to neutralize the particles and create agglomerates (Butler et al., 2011).

It has the capacity to remove a variety of pollutants from different types of wastewater, notably inorganic and organic contaminants. The pH, electrode, operating period, and current density are only a few of the variables that affect how well the electrocoagulation process works. Electrode passivation and energy use are the two key issues with the electrocoagulation method. Comparing electrocoagulation to other popular techniques provides benefits including lowering energy use and operating costs (Shahedi et al., 2020). Although electrocoagulation is a developing technique that is now used successfully to treat wastewater, the insufficient scientific knowledge on the intricate physical and chemical processes associated is restricting future design and slowing advancement (Mollah et al., 2001).

1.6.2 Electrodialysis

An applied potential difference is used as a driving factor in the membrane separation procedure known as electrodialysis to encourage ionic dissociation in aqueous medium. The method was first created to make drinking water from saltwater. However, because of its advantages, electrodialysis is being used more and more in the treatment of industrial wastewaters. The procedure is regarded as being clean since it makes it possible to recover materials and reclaim water. The development of electrodialysis as it is used to clean technologies in the electroplating business (Scarazzato et al., 2017). During electrodialysis, two streams are created that have different concentrations and flow in separate pockets that are divided into separate ones by cation-exchange membranes and anion-exchange membranes. Electrodialysis may be cost-effective because of the advantages it offers in obtaining high specificity and material recovery as well as the removed or decreased need for reagents. The development of ion exchange membranes, which increases water recovery and operates without the need of reagents, phase shifts, or reactions, is what makes electrodialysis possible. Without using energy sources or chemical cleansers, these benefits help the environment (Gurreri et al., 2020; Al-Amshawee et al., 2020). The production of extremely clean water, treatment of wastewater, the production of bases and acids by hydrogen production, and the transformation of chemical energy into electrical energy through various process combinations are now all possible uses for electrodialysis technology. Electro-adsorption, ion exchange, electrolysis, and microbial fuel cells technology are all used into the process integration to increase the effectiveness of pollutants removal from wastewaters or effluent (Moon and Yun, 2014).

1.6.3 Electrodeionization

Ionic separation system called electrodeionization first appeared about 50 years ago. Its research and use have been hampered by a poor understanding of its operational kinetics, which was initially used to separate metallic species from radioactive wastewater. The explanation of intricate operational mechanisms has been the focus of steadily expanding research efforts, allowing for the expansion of its applicability to various sectors. As of now, electrodeionization has excelled as a very effective, ecologically friendly process of purification, separation, and concentration. New materials are constantly being created to advance and mature this technology, which might have huge benefits for the environment and the global economy (Alvarado and Chen, 2014). The primary technical factors affecting an electrodeionization module's efficiency are its temperature, TDS, rate of flow, and current strength in the concentrated and diluted sections (Arar et al., 2014). The electrodeionization methodology has several key benefits over the other traditional inorganic separation methods, including longer ion-exchange resin life especially when compared to ion-exchange approach, prolonged membrane life contrast to membrane technology such as reverse osmosis/ultrafiltration, contaminant-free water treatment, easier ion-exchange resin process of restoration, and highly pure gear pollution which can be screened (Kumar et al., 2022). In terms of purifying, isolating, and concentrating, electrodeionization has proven to be an excellent, environmentally friendly method. In order to advance and develop this technology, new compounds are constantly being developed, and this might have enormous positive effects on the environment and the economy globally. In the current method, membranes have known constraints due to concentration polarization, membrane fouling, and scaling (Rathi et al., 2022).

1.7 Conclusion

The chapter focused on the current water demand and its reduction strategy. It also examined a variety of pollutants, including dyes, toxic metals, emerging contaminants, and others present in the wastewater. It also covered water-energy nexus and water sustainability. It also gives an overview of the most water treatment techniques such as adsorption, membrane technology, biological methods, and coagulation and flocculation. Also, it reviewed different electrochemical methods for wastewater treatment along with its advantage and disadvantage. In comparison with other conventional inorganic extraction methods, the electrodeionization process has a few significant advantages, such as a lengthier resin life, a simplified resin fast recovery, contaminant-free treatment of wastewater, a lengthier membrane life, and the ability to filter highly pure targeted pollutants.

References

Ahmad, S.Z.N., Salleh, W.N.W., Ismail, A.F., Yusof, N., Yusop, M.Z.M., Aziz, F., 2020. Adsorptive removal of heavy metal ions using graphene-based nanomaterials: toxicity, roles of functional groups and mechanisms. Chemosphere 248, 126008.

Al-Amshawee, S., Yunus, M.Y.B.M., Azoddein, A.A.M., Hassell, D.G., Dakhil, I.H., Hasan, H.A., 2020. Electrodialysis desalination for water and wastewater: a review. Chemical Engineering Journal 380, 122231.

Albatrni, H., Qiblawey, H., El-Naas, M.H., 2021. Comparative study between adsorption and membrane technologies for the removal of mercury. Separation and Purification Technology 257, 117833.

Alharbi, O.M., Khattab, R.A., Ali, I., 2018. Health and environmental effects of persistent organic pollutants. Journal of Molecular Liquids 263, 442−453.

Alvarado, L., Chen, A., 2014. Electrodeionization: principles, strategies and applications. Electrochimica Acta 132, 583−597.

Amin, M.T., Alazba, A.A., Manzoor, U., 2014. A review of removal of pollutants from water/wastewater using different types of nanomaterials. Advances in Materials Science and Engineering 2014.

Arar, Ö., Yüksel, Ü., Kabay, N., Yüksel, M., 2014. Various applications of electrodeionization (EDI) method for water treatment—a short review. Desalination 342, 16−22.

Azimi, A., Azari, A., Rezakazemi, M., Ansarpour, M., 2017. Removal of heavy metals from industrial wastewaters: a review. ChemBioEng Reviews 4 (1), 37−59.

Bafana, A., Devi, S.S., Chakrabarti, T., 2011. Azo dyes: past, present and the future. Environmental Reviews 19 (NA), 350−371.

Bhargava, A., 2016. Physico-chemical waste water treatment technologies: an overview. International Journal of Research in Education and Science 4 (5), 5308−5319.

Bolong, N., Ismail, A.F., Salim, M.R., Matsuura, T., 2009. A review of the effects of emerging contaminants in wastewater and options for their removal. Desalination 239 (1−3), 229−246.

Bulut, Y., 2007. Removal of heavy metals from aqueous solution by sawdust adsorption. Journal of environmental sciences 19 (2), 160−166.

Butler, E., Hung, Y.T., Yeh, R.Y.L., Suleiman Al Ahmad, M., 2011. Electrocoagulation in wastewater treatment. Water 3 (2), 495−525.

Castro-Muñoz, R., González-Melgoza, L.L., García-Depraect, O., 2021. Ongoing progress on novel nanocomposite membranes for the separation of heavy metals from contaminated water. Chemosphere 270, 129421.

Chen, G., 2004. Electrochemical technologies in wastewater treatment. Separation and purification Technology 38 (1), 11−41.

Dai, J., Wu, S., Han, G., Weinberg, J., Xie, X., Wu, X., et al., 2018. Water-energy nexus: a review of methods and tools for macro-assessment. Applied Energy 210, 393−408.

Deshmukh, A., Boo, C., Karanikola, V., Lin, S., Straub, A.P., Tong, T., et al., 2018. Membrane distillation at the water-energy nexus: limits, opportunities, and challenges. Energy & Environmental Science 11 (5), 1177−1196.

Dhangar, K., Kumar, M., 2020. Tricks and tracks in removal of emerging contaminants from the wastewater through hybrid treatment systems: a review. Science of the Total Environment 738, 140320.

Ding, T., Liang, L., Zhou, K., Yang, M., Wei, Y., 2020. Water-energy nexus: the origin, development and prospect. Ecological Modelling 419, 108943.

Duong, K., Saphores, J.D.M., 2015. Obstacles to wastewater reuse: an overview. Wiley Interdisciplinary Reviews: Water 2 (3), 199−214.

Fu, F., Wang, Q., 2011. Removal of heavy metal ions from wastewaters: a review. Journal of Environmental Management 92 (3), 407−418.

Fu, Z., Xi, S., 2020. The effects of heavy metals on human metabolism. Toxicology Mechanisms and Methods 30 (3), 167−176.

Gao, Q., Xu, J., Bu, X.H., 2019. Recent advances about metal−organic frameworks in the removal of pollutants from wastewater. Coordination Chemistry Reviews 378, 17−31.

Garcia, X., Pargament, D., 2015. Reusing wastewater to cope with water scarcity: economic, social and environmental considerations for decision-making. Resources, Conservation and Recycling 101, 154−166.

Gil, A., Assis, F.C.C., Albeniz, S., Korili, S.A., 2011. Removal of dyes from wastewaters by adsorption on pillared clays. Chemical Engineering Journal 168 (3), 1032−1040.

Gil, A., Taoufik, N., García, A.M., Korili, S.A., 2018. Comparative removal of emerging contaminants from aqueous solution by adsorption on an activated carbon. Environmental Technology .

Gupta, K., Joshi, P., Gusain, R., Khatri, O.P., 2021. Recent advances in adsorptive removal of heavy metal and metalloid ions by metal oxide-based nanomaterials. Coordination Chemistry Reviews 445, 214100.

Gupta, V.K., Ali, I., Saleh, T.A., Nayak, A., Agarwal, S., 2012. Chemical treatment technologies for waste-water recycling—an overview. RSC Advances 2 (16), 6380−6388.

Gurreri, L., Tamburini, A., Cipollina, A., Micale, G., 2020. Electrodialysis applications in wastewater treatment for environmental protection and resources recovery: a systematic review on progress and perspectives. Membranes 10 (7), 146.

Hamiche, A.M., Stambouli, A.B., Flazi, S., 2016. A review of the water-energy nexus. Renewable and Sustainable Energy Reviews 65, 319−331.

Hassaan, M.A., El Nemr, A., Hassaan, A., 2017. Health and environmental impacts of dyes: mini review. American Journal of Environmental Science and Engineering 1 (3), 64−67.

Jaiyeola, A.T., Bwapwa, J.K., 2016. Treatment technology for brewery wastewater in a water-scarce country: a review. South African Journal of Science 112 (3−4), 1−8.

Javaid, A., Bajwa, R., Shafique, U., Anwar, J., 2011. Removal of heavy metals by adsorption on Pleurotus ostreatus. Biomass and Bioenergy 35 (5), 1675−1682.

Kadirvelu, K., Thamaraiselvi, K., Namasivayam, C., 2001. Removal of heavy metals from industrial wastewaters by adsorption onto activated carbon prepared from an agricultural solid waste. Bioresource Technology 76 (1), 63−65.

Katheresan, V., Kansedo, J., Lau, S.Y., 2018. Efficiency of various recent wastewater dye removal methods: a review. Journal of Environmental Chemical Engineering 6 (4), 4676−4697.

Kumar, P.S., Varsha, M., Rathi, B.S., Rangasamy, G., 2022. Electrodeionization: fundamentals, methods and applications. Environmental Research 114756.

Lee, K.E., Morad, N., Teng, T.T., Poh, B.T., 2012. Development, characterization and the application of hybrid materials in coagulation/flocculation of wastewater: a review. Chemical Engineering Journal 203, 370−386.

Liu, L., Gao, Z.Y., Su, X.P., Chen, X., Jiang, L., Yao, J.M., 2015. Adsorption removal of dyes from single and binary solutions using a cellulose-based bioadsorbent. ACS Sustainable Chemistry & Engineering 3 (3), 432−442.

Mancosu, N., Snyder, R.L., Kyriakakis, G., Spano, D., 2015. Water scarcity and future challenges for food production. Water 7 (3), 975−992.

McKay, G., Porter, J.F., Prasad, G.R., 1999. The removal of dye colours from aqueous solutions by adsorption on low-cost materials. Water, Air, and Soil Pollution 114 (3), 423−438.

Mollah, M.Y.A., Schennach, R., Parga, J.R., Cocke, D.L., 2001. Electrocoagulation (EC)—science and applications. Journal of Hazardous Materials 84 (1), 29−41.

Mook, W.T., Aroua, M.K., Issabayeva, G., 2014. Prospective applications of renewable energy based electrochemical systems in wastewater treatment: a review. Renewable and Sustainable Energy Reviews 38, 36−46.

Moon, S.H., Yun, S.H., 2014. Process integration of electrodialysis for a cleaner environment. Current Opinion in Chemical Engineering 4, 25−31.

Moradihamedani, P., 2022. Recent advances in dye removal from wastewater by membrane technology: a review. Polymer Bulletin 79 (4), 2603−2631.

Muddemann, T., Haupt, D., Sievers, M., Kunz, U., 2019. Electrochemical reactors for wastewater treatment. ChemBioEng Reviews 6 (5), 142−156.

Noguera-Oviedo, K., Aga, D.S., 2016. Lessons learned from more than two decades of research on emerging contaminants in the environment. Journal of Hazardous Materials 316, 242—251.

Obotey Ezugbe, E., Rathilal, S., 2020. Membrane technologies in wastewater treatment: a review. Membranes 10 (5), 89.

Passow, H., Rothstein, A., Clarkson, T.W., 1961. The general pharmacology of the heavy metals. Pharmacological Reviews 13.

Paździor, K., Bilińska, L., Ledakowicz, S., 2019. A review of the existing and emerging technologies in the combination of AOPs and biological processes in industrial textile wastewater treatment. Chemical Engineering Journal 376, 120597.

Postel, S.L., 2000. Entering an era of water scarcity: the challenges ahead. Ecological Applications 10 (4), 941—948.

Prakash, N.B., Sockan, V., Jayakaran, P., 2014. Waste water treatment by coagulation and flocculation. International Journal of Engineering Science and Innovative Technology 3 (2), 479—484.

Qing Li, Q., Loganath, A., Seng Chong, Y., Tan, J., Philip Obbard, J., 2006. Persistent organic pollutants and adverse health effects in humans. Journal of Toxicology and Environmental Health, Part A 69 (21), 1987—2005.

Rashid, R., Shafiq, I., Akhter, P., Iqbal, M.J., Hussain, M., 2021. A state-of-the-art review on wastewater treatment techniques: the effectiveness of adsorption method. Environmental Science and Pollution Research 28 (8), 9050—9066.

Rathi, B.S., Kumar, P.S., Parthiban, R., 2022. A review on recent advances in electrodeionization for various environmental applications. Chemosphere 289, 133223.

Rathi, B.S., Kumar, P.S., Show, P.L., 2021. A review on effective removal of emerging contaminants from aquatic systems: current trends and scope for further research. Journal of Hazardous Materials 409, 124413.

Russell, S., Fielding, K., 2010. Water demand management research: a psychological perspective. Water Resources Research 46 (5).

Saleh, T.A., Mustaqeem, M., Khaled, M., 2022. Water treatment technologies in removing heavy metal ions from wastewater: a review. Environmental Nanotechnology, Monitoring & Management 17, 100617.

Scarazzato, T., Panossian, Z., Tenório, J.A.S., Pérez-Herranz, V., Espinosa, D.C.R., 2017. A review of cleaner production in electroplating industries using electrodialysis. Journal of Cleaner Production 168, 1590—1602.

Schleich, J., Hillenbrand, T., 2009. Determinants of residential water demand in Germany. Ecological Economics 68 (6), 1756—1769.

Seckler, D., Barker, R., Amarasinghe, U., 1999. Water scarcity in the twenty-first century. International Journal of Water Resources Development 15 (1—2), 29—42.

Shahedi, A., Darban, A.K., Taghipour, F., Jamshidi-Zanjani, A., 2020. A review on industrial wastewater treatment via electrocoagulation processes. Current Opinion in Electrochemistry 22, 154—169.

Shahid, M.K., Kashif, A., Fuwad, A., Choi, Y., 2021. Current advances in treatment technologies for removal of emerging contaminants from water—a critical review. Coordination Chemistry Reviews 442, 213993.

Shahrokhi-Shahraki, R., Benally, C., El-Din, M.G., Park, J., 2021. High efficiency removal of heavy metals using tire-derived activated carbon vs commercial activated carbon: insights into the adsorption mechanisms. Chemosphere 264, 128455.

Shammas, N.K., 2005. Coagulation and flocculation. Physicochemical Treatment Processes. Humana Press, pp. 103—139.

Shrestha, R., Ban, S., Devkota, S., Sharma, S., Joshi, R., Tiwari, A.P., et al., 2021. Technological trends in heavy metals removal from industrial wastewater: a review. Journal of Environmental Chemical Engineering 9 (4), 105688.

Singh, R., Gautam, N., Mishra, A., Gupta, R., 2011. Heavy metals and living systems: an overview. Indian Journal of Pharmacology 43 (3), 246.

Stathatou, P.M, Gad, F.K., Kampragou, E., Grigoropoulou, H., Assimacopoulos, D., 2015. Treated wastewater reuse potential: mitigating water scarcity problems in the Aegean islands. Desalination and Water Treatment 53 (12), 3272−3282.

Taoufik, N., Elmchaouri, A., Anouar, F., Korili, S.A., Gil, A., 2019. Improvement of the adsorption properties of an activated carbon coated by titanium dioxide for the removal of emerging contaminants. Journal of Water Process Engineering 31, 100876.

Ungureanu, N., Vlǎduţ, V., Voicu, G., 2020. Water scarcity and wastewater reuse in crop irrigation. Sustainability 12 (21), 9055.

Urban, J.J., 2017. Emerging scientific and engineering opportunities within the water-energy nexus. Joule 1 (4), 665−688.

Vakilifard, N., Anda, M., Bahri, P.A., Ho, G., 2018. The role of water-energy nexus in optimising water supply systems−review of techniques and approaches. Renewable and Sustainable Energy Reviews 82, 1424−1432.

Wang, J., Zhuang, S., 2017. Removal of various pollutants from water and wastewater by modified chitosan adsorbents. Critical Reviews in Environmental Science and Technology 47 (23), 2331−2386.

Wang, X.C., Jiang, P., Yang, L., Van Fan, Y., Klemeš, J.J., Wang, Y., 2021. Extended water-energy nexus contribution to environmentally-related sustainable development goals. Renewable and Sustainable Energy Reviews 150, 111485.

Xiang, H., Min, X., Tang, C.J., Sillanpää, M., Zhao, F., 2022. Recent advances in membrane filtration for heavy metal removal from wastewater: a mini review. Journal of Water Process Engineering 49, 103023.

Xu, W., Wang, X., Cai, Z., 2013. Analytical chemistry of the persistent organic pollutants identified in the Stockholm Convention: a review. Analytica Chimica Acta 790, 1−13.

Yadav, S., Kamsonlian, S., 2022. A review of electrochemical methods for treatment of wastewater. Materials Today: Proceedings .

Yildiz, B.S., 2012. Water and wastewater treatment: biological processes. Metropolitan Sustainability. Woodhead Publishing, pp. 406−428.

Zhou, Y., Lu, J., Zhou, Y., Liu, Y., 2019. Recent advances for dyes removal using novel adsorbents: a review. Environmental Pollution 252, 352−365.

Zubaidi, S.L., Ortega-Martorell, S., Al-Bugharbee, H., Olier, I., Hashim, K.S., Gharghan, S.K., et al., 2020. Urban water demand prediction for a city that suffers from climate change and population growth: Gauteng province case study. Water 12 (7), 1885.

CHAPTER TWO

Technology overview of electrodeionization

2.1 Introduction

Governments and academic institutions are paying an increasing amount of attention to the recovery of valuable materials from wastewater due to the rapid increase in the global population. Over the last three decades, substantial research into electrochemical technology for wastewater purification has been conducted. Nevertheless, the use of these techniques to recover resources from wastewater has only recently drawn a small amount of interest (Liu et al., 2021). For more than 50 years, saline water sources have been converted into drinkable water using electrodialysis (Strathmann 2010a,b; Tian et al., 2020). Metals from acid wastewaters can be concentrated, separated, and extracted in a specific way using this technique. ED looks to be a feasible technology with great growth potential. It is capable of dealing with low metal concentrations. Precipitation in the setup, high power consumption, waste disposal, low selectivity, and system scaling up are some of typical issues (Juve et al., 2022).

Ion exchange is the broad term for each time an ion is taken out of an aqueous medium and exchanged by another ionic species. An ion exchanger is a substance that is not soluble in water and can exchange some of its particles like charged ions found in a solution (Gu et al., 2019). Because the substance was inexpensive and resin renewal capability, it is reasonably affordable and has an increased treatment capacity, quick kinetics, and high extraction efficiencies. Some of the demerits are fouling, increased sensitivity to pH changes, decreased binding affinity from other chemicals, and secondary waste which are all effects of resin renewal (Kumar and Jain, 2013).

Ion-exchange resins (IERs) and ion-exchange membranes (IEMs) are used together in the hybrid extraction technique known as electrodeionization (EDI), commonly referred to as continuous deionization. In order to extract or reclaim water-based ions, EDI has drawn growing interest (Arar et al., 2014). Diverse species accumulation and eradication from effluent stream has a variety of uses for EDI. It replaces the dangerous chemicals that are often used to renew resins with electric power, eliminating the effluent connected with resin renewal. Electrodeionization removes various issues with IER beds in the water treatment procedure, most noticeably ion discharge as the beds wear out (Rathi et al., 2022). However, a technique known as EDI was developed that is incredibly successful in removing ionic compounds from contaminated waterways. High-purity water may be made via a process called continuous EDI. Along with producing purified water, the process also has innovative wastewater treatment technologies that make it easier to get rid of heavy metals, radioactive pollutants, toxic chemicals, and other dangerous impurities. Innovative materials have been created to progress and enhance this method, which could have a hugely positive impact on the environment and the economy globally (Kumar et al., 2022).

This chapter's goal is to provide a comprehensive analysis of the ion exchange and electrodialysis processes. It also examined the workings, design, benefits, and drawbacks of EDI, ion exchange, and electrodialysis. Basics of electrochemistry, the current–voltage relationship in EDI, the Nernst equation, the Donnan potential, electrical resistance, limiting current density, and the transport mechanism in IERs are just a few of the topics covered in fundamentals of EDI.

2.2 Electrodialysis

Electrodialysis is the main electrical and chemical technique utilized in industry for way to solve demineralization. In an electrochemical purification process called electrodialysis, ionic compounds are moved from one solution to another by passing through one or more partially permeable membranes under the impact of a direct current (DC). The two main cell types that could be employed for electrodialysis are multimembrane cells for diluting and concentrating purposes and electrolysis cells for redox processes (Bazinet et al., 1998).

2.2.1 Principle of electrodialysis unit

A DC voltage is used in the electrochemical separation procedure known as electrodialysis to move ions via IEMs. Using a driving force, the process transfers positively and negatively charged ions from the feedwater through cathodes and anodes to a flow of saturated effluent, leading in a stream that is more dilute. By transferring the ions from the saltwater across an electrically charged semipermeable IEM, electrodialysis preferentially eliminates dissolved particles depending on their electric charges (Valero et al., 2011). The electrodialysis cell that uses a voltage was applied across cathode and anode electrodes that pass through an IEM to extract ionic particles from uncharged materials in an aqueous medium. Consequently, it is shown that electrodialysis is an electrically powered process.

Particularly, cations and anions move toward the cathode and anode, respectively. IEM acts as perfect preferential boundaries to ions, allowing anions to readily travel via nearest anion–exchange membrane (AEM) despite being prevented by the nearest cation–exchange membrane (CEM) (Dammak et al., 2021). In contrast, cations pass via CEM and are inhibited by AEM, which causes the salt concentration of the dilute compartments to decrease and the concentrate one to become more enriched. The induced electrical potential causes a reduction process at the cathode when a feed stream penetrates an electrodialysis stack, producing hydroxyl radicals in the cationic chamber. A cation from the feed chamber will move across a CEM to keep electroneutrality. Likewise, to how protons are produced in the anion chamber by oxidation at the anode, anions are retrieved from the feed as it passes through CEM. In order to produce concentrated and diluted effluents, salts are thus extracted from the feed stream and added to the concentrated chamber. Table 2.1 depicts the eradication of pollutants utilizing electrodialysis along with operating condition. Examples of cationic substances that are moving via the CEM and into the cathode chamber include ammonia, potassium, and sodium. AEM transfers anionic substances such as phosphate, sulfate, and chlorine to the anode portion, maintaining the matrix's electric equilibrium (Al-Amshawee et al., 2020).

2.2.2 Construction of electrodialysis unit

Electrodialysis is an electromechanical separation technique that extracts ions from electrolytes in addition to hardness and organic material. It uses IEMs within an electric field to facilitate ionic separation (Mohammadi et al., 2021). A conventional electrodialysis unit is made up of a group of AEM and CEM that are ion selective, alternate compartments for

Table 2.1 Removal of contaminants using electrodialysis along with operating condition.

Contaminants	Electrode	Membrane	Operating condition	Removal percentage	References
Arsenic	Platinum coated electrodes	AEM AR204 SZRA B02249C CEM CR67 HUY N12116B	Temperature—30°C pH—2 Contact time—16-h Initial concentration—0.016 mg/g	30	Sun and Ottosen (2012)
Cadmium	Carbon and iron	—	Temperature—50°C Agitation speed —100 rpm Contact time—8-h Initial concentration—163 μg/L	78.5	Sivakumar et al. (2014)
Cadmium	Anode and cathode	AEM	Temperature—30°C pH—3.7 Contact time—24-h Initial concentration—20 mg/g	31	Ebbers et al. (2015)
Cadmium	Platinum coated electrodes	AEM AR204 SZRA B02249C CEM CR67 HUY N12116B	Temperature—30°C pH—2 Contact time—16-h Initial concentration—0.212 mg/g	17	Sun and Ottosen (2012)
Chromium	Anode and cathode	AEM	Temperature—30°C pH—3.7 Contact time—24-h Initial concentration—30 mg/g	6	Ebbers et al. (2015)

Chromium	Platinized titanium	AEM CEM	Temperature—30°C Applied voltage—3.3 V Contact time—1 week Initial concentration—1.06 mg/g	86	Ottosen et al. (2003)
Copper	Anode and cathode	AEM	Temperature—30°C pH—3.7 Contact time—24-h Initial concentration—60 mg/g	22	Ebbers et al. (2015)
Copper	Platinum coated electrodes	AEM AR204 SZRA B02249C CEM CR67 HUY N12116B	Temperature—30°C pH—2 Contact time—16-h Initial concentration—1.5 mg/g	54	Sun and Ottosen (2012)
Copper	Platinized titanium	AEM and CEM	Temperature—30°C Applied voltage—3.3 V Contact time—1 week Initial concentration—3.3 mg/g	94	Ottosen et al. (2003)
Lead	Platinized titanium	AEM and CEM	Temperature—30°C Agitation speed—1300 rpm Contact time—10 weeks Initial concentration—400 μg/g	8	Pedersen et al. (2005)
Lead	Platinized titanium	AEM and CEM	Temperature—30°C Agitation speed—1300 rpm Contact time—3 weeks Initial concentration—400 μg/g	2.5	Pedersen et al. (2005)

(*Continued*)

Table 2.1 (Continued)

Contaminants	Electrode	Membrane	Operating condition	Removal percentage	References
Lead	Anode and cathode	AEM	Temperature—30°C pH—3.7 Contact time—24-h Initial concentration—20 mg/g	1	Ebbers et al. (2015)
Lead	Platinum coated electrodes	Heterogeneous CEMs	Temperature—30°C Applied voltage—15 V Contact time—1-h Initial concentration—50 mg/L	99.9	Nemati et al. (2017)
Nickel	Anode and cathode	AEM	Temperature—30°C pH—3.7 Contact time—24-h Initial concentration—30 mg/g	56	Ebbers et al. (2015)
Nickel	Platinum coated electrodes	Heterogeneous CEMs	Temperature—30°C Applied voltage—3.3 V Contact time—1 week Initial concentration—50 ppm	96.9	Nemati et al. (2017)
Potassium	Platinum coated electrodes	Heterogeneous CEMs	Temperature—30°C Applied voltage—15 V Contact time—1-h Initial concentration—50 ppm	99.9	Nemati et al. (2017)
Tin	Carbon and iron	—	Temperature—50°C Agitation speed—100 rpm Contact time—8-h Initial concentration—122 μg/L	67.3	Sivakumar et al. (2014)

Zinc	Platinized titanium	AEM CEM.	Temperature—30°C Agitation speed—1300 rpm Contact time—10 weeks Initial concentration—22.65 mg/g	73	Pedersen et al. (2005)
Zinc	Platinized titanium	AEM CEM.	Temperature—30°C Agitation speed—1300 rpm Contact time—3 weeks Initial concentration—22.65 mg/g	24	Pedersen et al. (2005)
Zinc	Anode and cathode	AEM	Temperature—30°C pH—3.7 Contact time—24-h Initial concentration—100 mg/g	85	Ebbers et al. (2015)

AEM, Anion-exchange membrane; *CEM*, cation-exchange membrane.

concentrated and diluted solution, and electrode chambers at the ends of each set of membranes. Numerous CEM and AEM are situated in between the anode and cathode electrodes within the electrodialysis stack. To divide the IEM and generate concentrated and diluted chambers, a spacers' gasket is used inside the electrodialysis stack (Jaroszek and Dydo, 2016). When it comes to the spacers' construction, there are two key ideas. They are the tortuous path idea and the sheet-flow spacer idea. The vertical positioning of the chambers and the compressed length of the flow pathway are the main differences between the sheet-flow separator and the tortuous pathway flow separator. The liquid's rate of flow in the chambers made of two membranes and a spacer is $2-4$ cm/s, and there is a $0.2-0.4$ bar pressure drop. The membrane spacers have a long helical cut-out that creates a longer, narrower tunnel for the fluid flowing and are horizontally positioned in the tortuous route flow stack. Improvements in concentration polarization control and higher limiting current densities are still feasible despite the stack's extremely high feed flowrate of $6-12$ cm/s because of the drop in pressure in the feed flowing fluid, which is quite significant at $1-2$ bar (Strathmann, 2010a,b). Fig. 2.1 depicts the electrodialysis cell.

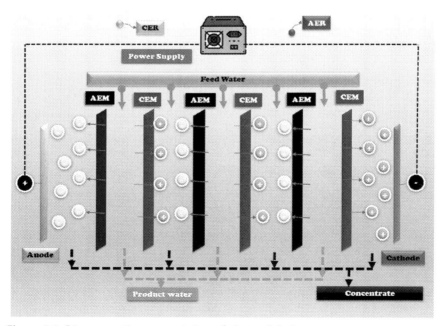

Figure 2.1 Diagrammatic representation of electrodialysis.

Parallel membranes keep the electrodes roughly apart. As a hindrance to the movement of nutrients, membranes either stop or allow ions to flow based on their electrical current. Via electrode chambers, also known as electrode rinse containers, the electrolyte solution is exchanged (Campione et al., 2018). The characteristics of the membranes have a significant role in determining how effective electrodialysis is. However, the process and system construction, which establishes the current utilization, limiting current density, concentration polarization, and so on, also has an impact. Therefore, the total effectiveness and costs of electrodialysis are significantly influenced by the construction process (Wolfaardt et al., 2021).

2.2.3 Drawbacks

Although all of the clear benefits, electrodialysis has certain problems that must be fixed. When scaling up the electrodialysis technique, the expenses are typically the biggest obstacle. The membrane stack and its initial outlay are still expensive. The high energy demand is mostly brought on by the high activation overpotential needed to run the unit owing to the resistance of the cell's constituent parts (Juve et al., 2022). These techniques also suffer from the limitations associated with membranes, such as blockage, fouling, renewability, and selectivity, which add to the high cost of operation. Due to the absence of efficient selective membranes, the specificity of the membranes may be a disadvantage relying on the metal mixture's separation ability (Nazif et al., 2022; Sun et al., 2020). Electrodialysis is a technique that is still under progress but has enormous growth potential. To improve the selectivity and/or lower the electron flow resistance, advanced materials and cell components are being developed. By optimizing both the materials and the operating parameters, the energy needed to run the unit and its efficiency may be reduced. Additionally, solutions to issues including the production of metal concentrates, the scaling of inorganic solids, and ability to scale restrictions are needed (Mir and Bicer, 2021).

2.3 Ion exchanger

Ion-exchange technology has been receiving gradually increased recognition in a number of industries for a number of years. This technique is widely used to clean solutions by electrostatic interaction sorbing the dissolved substances from the solution using ion-exchange materials in a variety of physical forms (Pang et al., 2021). To refill the removed ions, equal amounts of new ions with the same charges are introduced to the solutions. Ion-exchange processes allow for the separation and purification of other ions or the eradication of particular ions from a fluid. It is feasible to differentiate between complete deionization and the elimination of certain ionic pollutants as a consequence. The solution's constitution and the required level of filtration mostly determine which option is best (Nasef and Ujang, 2012).

2.3.1 Principle of ion exchanger

A procedure known as ion exchange involves replacing counter-ions in a solid matrix with cations and anions that are soluble in an electrolysis cell. Counter-ions exchange with soluble salts of the same sign and charge on a stoichiometry equitable level. Regeneration can stop the process, and the stationary phase ion exchanger's structural stability is not compromised (Wang and Wan, 2015). IERs are used in water purification plants to replace one ion with a different one in order to accomplish the goal of demineralization. IERs come in two different varieties: anion-exchange resins and cation resins. Its components may be divided into several groups and systems depending on whether they are strong base cation, weak acid anion, or weak base anion (Muhammad et al., 2019). The kind of exchangers does not really matter because resins are particularly susceptible to the fouling phenomenon, which is brought on by the continuous presence of biomass in untreated water. It is crucial to utilize a separate portion to eliminate any form of suspended materials that may be present. Even before liquid goes through any treatment processes, it ought to be done. Additionally, soluble organics must to be eliminated if at all feasible to reduce the burden on the ion-exchange device (Kansara et al., 2016).

Although there are many parallels between ion-exchange phenomena and adsorption mechanisms, there are also a few notable distinctions. The chemical species taken into consideration in this kind of process are ions, which are not removed from solutions but rather replaced by ions connected to the solid substrate by electrostatic attraction to achieve electrical neutrality. As a result, there seem to be two ionic fluxes: one entering the ion-exchange particle and the other leaving the resin particles in the opposite direction. Adsorbent resins have been used to address the majority of ideas and models, and in many real-world applications, sorption and ion-exchange processes are not separated (Ma et al., 2019). This is due to the fact that the problem is simpler if the resins are devoid of ligands. The concept that adsorption also takes place onto the resin matrix as a consequence of hydrophobic interactions when ionic organic molecules are exchanged is one feature that has so far received less attention. This increases the complexity of evaluating ion-exchange data greatly (Kammerer et al., 2011).

Ion exchangers are solid materials with the ability to take in positive particles from solutions and then release an equal amount of other ions returned into the source. The structural nature of the materials determines their capacity for exchanging ions. The exchanger is made out of a grid with either a $+$ve or $-$ve excess charge. This excess charge is localised in functional groups or at specific sites in the solid structure. Counter-ions, which may move inside the matrix's empty space and could be substituted by other ions of the same charge, balance out the matrix's charge. The exchangers' pores can hold on to counter-ions and solvent. Depending on the composition of the counter-ions, this may result in swelling. The open spaces of varying sizes and shapes found in an exchanger are known as pores. As a result, the exchangers display an irregularly sized three-dimensional network of channels. Besides the solvent, several electrolytes can also enter the exchanger. Co-ions are extra counter-ions with a similar charge to the stationary ions as a consequence (Berrios et al., 2013).

2.3.2 Construction of ion exchanger

IERs are insoluble polymers with basic or acidic ligand groups that may exchange opposing ions in the surrounding aqueous medium. Adsorption phenomena known as ion exchange use an electrostatic mechanism. Ions are attracted by electrostatic interaction to the functional groups of the IER and are then replaced by the attracted ions in a 1:1 charge ratio. The resins are made from an substrate's biopolymer spine and come in minute beats that are often white or yellowish ($1-2$ mm). The majority of the beats are pores, offering a large surface area that helps to preserve electroneutrality. Each and every ion exchanger has a set ionic group that is countered by an additional ion (Ijanu et al., 2020).

2.3.3 Drawbacks

High disposal costs for currents in the reduction phase. In spite of heavy pollution, resin lifetime is poor. When organic pollutants are present, the majority of resins are contaminated. Its ability to eliminate all ionic substance from solutions is what makes it advantageous. It is quite effective, and a wide range of resins are available that are simple to reuse or recycle. IER technology's drawback is that it must first go through a prefiltration procedure in order to be employed. Costly large equipment is possible. Media regeneration could need for additional hardware. Inorganic ion exchangers are challenging to move via pipes. Only a small portion of the bed particle is actively engaged in ion exchange at any given moment (Esmaeili and Foroutan, 2015).

One of the least expensive and most extensively used cation resin regenerants is sulfuric acid. Some water sources have a high calcium content, which when combined with sulfuric acid, produces calcium sulfate. As a result, the resin becomes contaminated and scale deposits clog drain pipes. Oxidation of Fe^{2+} to Fe^{3+} takes place, and as a result, ferric hydroxide precipitates, clogging resin beds once more and preventing ion exchange. The most frequent reason for softening failure is iron clogging. On-site chemical cleaning is possible for both organic and iron fouled units; however, total impurity removal is uncommon, and resin efficiency often declines following fouling. Disinfectants such as formaldehyde can be employed to clean resin beds; however, heat and oxidizing disinfectants such as chlorine can harm the resins. Such feeds are often treated by being run through activated carbon, which effectively eliminates chlorine (Mikhaylin and Bazinet 2016; Kansara et al., 2016).

2.4 Electrodeionization

The process of EDI, which integrates electrodialysis with traditional ion exchange, is an established one that has been used for more than 30 years in the manufacturing of ultrapure water. The eradication of chemical regeneration is the main reason in its commercialization. Due to the expanding use of the EDI process, a very large plant was built as the commercial, cutting-edge water manufacturing facility for high-pressure boilers (Shen et al., 2022). More lately, the EDI system has discovered a variety of intriguing new uses in the biotechnology sector, treating wastewater, and other promising fields. The evolution of stack construction and arrangement are also starting to cause worry, along with continued expansion and expanded applications (Khoiruddin et al., 2014). Along with producing pure water, the process also has exciting wastewater treatment technologies that make it easier to get rid of heavy metals, nuclear contaminants, toxic chemicals, and other dangerous impurities. Innovative materials have been created to progress and enhance this method, which would have a hugely positive impact on the environment and the economy globally (Kumar et al., 2022).

2.4.1 Principle of electrodeionization

A hybrid technology called EDI is dependent on the use of two separate techniques, notably electrodialysis and ion exchange. In addition to create a platform for electrical activity and stop the concentration polarization phenomena, beds of IERs are added to the diluted chambers. The dynamic media, which gathers and releases the ionic species, facilitates transport of ions via two distinct compartments because of the used electric current (Ulusoy Erol et al., 2021). This starts mass transfer through the ion-exchange components. At locations inside the unit where cationic and anionic exchange materials are in connection, water dissociates concurrently, resulting in H^+ and OH^- that work to replenish the resin in place. The purpose of the IER is to reduce the cell's impedance, which rises as the intensity of the dilution fluid decreases (Alvarado and Chen, 2014). Table 2.2 gives the information about removal of contaminants using EDI along with operating condition.

It has two distinct functioning modes: electrical technique recuperation and effective movement. The effective movement procedure uses the IER to movement of ions across the surface of the IEM. Only technology that enables the parallel removal from both cations and anions in order to maintain electroneutrality may be suitable for such a technique of ion eradication (Jordan et al., 2020). The consistent recovery of resins by H^+ and OH^- ions by those engaged in the electrically powered water splitting technique serves as the definition of the recovery by electrical technique. Ideally, this dissociation takes place at the bipolar interface's ion-depleting chamber. In an EDI unit, the ion cleanup happens in two steps: (1) the distribution of cations and anion to the strong cation and anion exchangers, correspondingly; and (2) ionic conductance of a stationary

Table 2.2 Removal of contaminants using electrodeionization along with operating condition.

Contaminants	Resin	Membrane	Operating condition	Removal Percentage	References
Cobalt	Strong acid CER, AMBERLITE IRN77 a strong base AER, AMBERLITE IRN78	NEOSEPTA CMX CEM AMX AEM	Flow rate—5 mL/min Temperature—25°C Applied current density—17 mA/cm^2 Initial concentration—0.34-mM	99	Yeon et al. (2004)
Copper	D001 strongly acid styrene-divinylbenzene CER D201 strongly alkali styrene-divinylbenzene AER	Polyethylene heterogeneous AEM and CEM	Flow rate—3 L/h Temperature—25°C pH—4.28 Time—8-h Initial concentration—50 ppm	99.5	Feng et al. (2008)
Nickel	D072 strongly acidic D296 strongly basic resins	CEM and AEM sulfonic acid and quaternary ammonium groups	Flow rate—20 L/h Temperature—25°C pH—3 Time—10-h Initial concentration—50 ppm	97	Lu et al. (2014)
Nickel	D072 strongly acidic D296 strongly basic resins	CEM and AEM sulfonic acid and quaternary ammonium groups	Flow rate—5 L/h Temperature—25°C pH—3 Time—10-h Initial concentration—55 ppm	99	Lu et al. (2011)

Chromium	AER	CEM, Nafion117, DuPont Co	Flow rate—1.2 L/h Temperature—25°C pH—3 Time—2-h Initial concentration—50 ppm Applied current density—375 A/m^2	99	Jina et al. (2020)
Chromium	Amberlite IRA-67 Dowex Mac-3	CEM and AEM	Flow rate—12 mL/min Temperature—25°C pH—5 Time—1.3-h Initial concentration—100 ppm Applied current density—375 A/m^2	98	Alvarado et al. (2013)
Chromium	Gel-type strong-base resin	Heterogeneous AEM and CEM	Flow rate—1.2 L/h Temperature—25°C pH—5 Time—16-h Initial concentration—100 ppm Applied current density—50 A/m^2	89.5	Xing et al. (2009)

AEM, Anion-exchange membrane; *CEM*, cation-exchange membrane.

surface only at the edges of the membranes. While in multilayered bed devices, water divides largely at the surface of the membrane or at the resins, water electrolysis at either interfaces may lead to efficient resin recovery for mix bed frameworks (Rathi and Kumar, 2020; Otero et al., 2021).

2.4.2 Construction of electrodeionization

EDI units (Fig. 2.2) are electrochemical equipment powered by a DC external power source. Each EDI device is made up of the following five main parts: ion-exchange resin, CEM and AEM, and cathode and anode electrodes are the major components of the system (Lee et al., 2016). There are two electrodes on the terminals of several cell pairs. One dilute chamber and one concentrate chamber are included in cell pairs. Three operations are going on at once while electricity and water are being applied to the EDI unit; (1) ion movement, in which the ions are evacuated from the resin; (2) deionization, in which the water is cleaned by ion exchange; and (3) ongoing resin renewal (Tate, 2014).

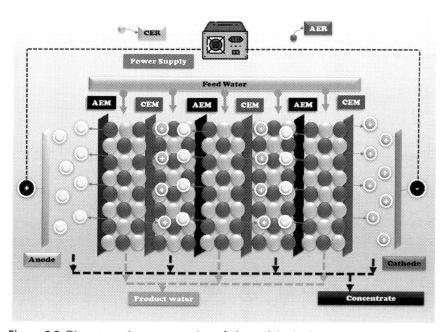

Figure 2.2 Diagrammatic representation of electrodeionization.

A cell couple's diluting compartments are the parts that include mixed bed IER for purifying or diluting ions from water. Water that has had its ions collected in concentrating compartments eventually becomes waste water. Both CER and AER are present in the dilution cells. Ion selective membranes divide the concentrating and diluting chambers. In terms of composition and charge, the membranes are comparable to the IER (Wenten et al., 2020). Only cations and anions may travel across AEM and CEM, respectively. IEM employed in EDI may not allow the passage of water or ions with opposing charges. Deionization is the process of removing ions, including positively and negatively charged cations and anions. Since they have lost one or more negatively charged ions, cations are positive-charged ions. Anions have a negative charge since they possess one extra electron or many (Hakim et al., 2020).

2.4.3 Merits of electrodeionization

EDI has gone a long way from its commercial start in 1987. EDI offers adaptable options for separations in the chemical, biological, food, and pharmaceutical industries since it can successfully remove a range of contaminants, as proven and discussed (Mistry et al., 2022). The deionization performance of EDI is always improving, and its single unit flow velocity capability is growing with every new generation, as is the characteristic of any advanced technology device of this period. The major technical factors that affect an EDI device's performance are its temperature, salinity, flow rate in the diluted and concentrated chambers, and current strength (Sarıçiçek et al., 2021). The past patterns in EDI systems include decreased labor and material costs, decreased difficulty, increased reliability, and decreased footprints (Zahakifar et al., 2020). The main running costs for EDI systems are replacing EDI modules, using electricity, treating water, and maintaining laboratory controls. These expenses are inversely correlated with system production and water supply quality (Chen et al., 2021).

The absence of regeneration and its accompanying harmful compounds, ensuring safety, and promoting worker health have all been mentioned as advantages of EDI. The membranes and resins are indestructible. Additionally, the product water's quality does not change over time. Continuously deionized water-requiring processes are anticipated and must have two separate units: one that provides clean water and the other that replenishes it. This duplexing ends up being costly and difficult. The size of conventional deionization systems is increased. Appreciably, since EDI is a continuous operation rather than a batch one, duplexing is no longer required. The outcome of EDI device footprints usually measures half as long as their conventional counterparts (Moon and Yun, 2014).

2.5 Fundamentals of electrodeionization

Operating factors such as fluid flow rate, current density, and feed intensity, EDI layer configuration, which is the thickness of the EDI chamber, and IER factors such as packing ratio, packing density, diameter, and accessible surface area all affect how well EDI separates materials. An enhanced EDI performance can be attained by designing such characteristics (Narayanan et al., 2009).

2.5.1 Electrochemistry

IERs are placed inside the ion-depleting or ion-concentrating chambers of an EDI system. The IERs improve ion mobility and can function as a platform for electrochemical processes such as the ionization of water to produce H^+ and OH^- ions (Arar et al., 2014). An instance of sodium chloride salt elimination may be used to illustrate the ion removal process in CER and AER. For the extraction of sodium chloride salt and the recovery of resin, the reaction pathway has been depicted (Rathi and Kumar, 2020).

2.5.2 Current−voltage relationship in electrodeionization

EDI operation produced the relationship between current density and applied voltage. There is no section of the EDI curve with a zero-gradient line, and the resistance tends to decrease. It implies that in the EDI method, the concentration polarization can be reduced. Additionally, until a voltage of 40 V, the increase in applied current obeys Ohm's law. Outside of that voltage, the applied current grows nonlinearly instead of linearly owing to the extra proton and hydroxyl ions in the dilute compartments (Ervan and Wenten, 2002; Song et al., 2005). The number of soluble ions carried increases with increasing current density. Three separate zones were seen in the velocity and current relation. First is referred to as the ion transport region, which is represented by a linear relationship between Volt and current. The movement of proton and hydroxyl ions from the water disintegration process at the bipolar boundary is what distinguishes the second zone as the water splitting zone, which is shown by a sharp rise in the velocity and current slope. The water splitting limitation zone is the third zone. The applied current region is created when water diffuses into the transitional zone and lowers the levels of hydrogen and hydroxyl ions (Hakim et al., 2020; Lee et al., 2017).

2.5.3 Donnan potential

The internal boundaries are in Donnan balance. The classic Donnan relations are satisfied by the component concentration at the membrane—solution interface. The Donnan potential is used to determine the electric potential rise at the internal borders. This sets limitations on every compartment's potential. For the internal borders, the current and ionic flow must remain continuous (Lu et al., 2016). The Donnan potential is defined by the following equation at the interface of the IEM and solution.

$$\Delta\varphi\frac{m}{D} = \varphi^m - \varphi^l = -\frac{RT}{z_i F}\ln\frac{C_m^i}{C_i^l}$$

which are met by the CEM with $I = H^+$ and Na^+ (m = cation membrane) and the anion-exchange membranes with $I = Cl^-$ (m = anion membrane). φ is the electric potential, D represents Donnan, C represents concentration, z_i is the charge, and F is the Faraday constant (Otero et al., 2021; Lu et al., 2010).

2.5.4 Electrical resistance

The relationship between the volt supplied and the current that flows thru a network determines its electrical resistance. The current or the electrons begin to move when the voltage difference is introduced to a device. The atoms and molecules are struck by the electrons while they are in motion. The passage of electrons or electrical currents is slowed down by collision or obstacle (Meaden, 2013). Raising the voltage and/or increasing the current requires lowering the component's electrical resistance. Numerous negative effects are brought by raising the voltage, including elevated safety hazards and diminished dependability. It was demonstrated that the hydrogen production reaction may be catalyzed by the use of strongly basic and acidic IEM, lowering the module resistance and enhancing weak ion removal. Reducing the resistance of the concentrated stream is another method for lowering the electrical resistance of the whole component. The concentrated stream is actually the module's restricting resistance when there is IER in the dilute cell. Increasing the water's conductivity will lower the concentrated stream's impedance (Gifford and Atnoor, 2000). It should be mentioned that the economics of an EDI system are determined by the expense of electrical energy and the initial cost of the EDI stack. The electrical resistance of the EDI stack and the input current density affect power consumption, while the membrane area affects capital costs. While the specific energy consumed rises according to current densities, the necessary membrane area reduces as input current density increases. The optimal cost of the EDI technique is estimated by the conflicting trends of energy and capital prices since cost of maintenance is relatively independent of applied current density. Therefore, it is essential to precisely adjust the membrane area and applied current density (Widiasa et al., 2004).

2.5.5 Limiting current density

When an electric current passes via an IEM, concentration polarization causes the intensity of salt on a desalination surface of the membrane to fall until it reaches zero at the limiting current density. Concentration polarization occurs on the desalination surface of an IEM boundary. Both ionic transportation in the boundary and solution movement in a desalination unit have an impact on the limiting current density. The distribution of solution flow in desalination cells affects the limited current density of an electrodialyzer (Tanaka, 2007). When the limiting current density is exceeded, the salt transfer from the transitional zone can no longer be balanced by the transfer into the region, resulting in a significant rise in membrane impedance and faster water dissociation (Strathmann et al., 1997).

The limiting current density beneath a flowing liquid in a desalination cell was understandable using the Nernst diffusion formula, which assumes that the layer of a boundary is equal to the thickness of a diffuse layer and illustrated by the feature of sodium chloride intensity and linear rate of a reaction mixture in a desalination cell. A spacer is commonly assumed to promote disruptions. In a desalination cell, however, it appears to diminish the limiting current density, produce dead zones between the separator and the membrane, and disturb the main laminar flow. To increase the

Technology overview of electrodeionization 47

limiting current density, the solution rate and Reynolds number must be increased, leading in disturbance (Tanaka, 2005).

2.5.6 Nernst equation

The Nernst equation connects the conventional cell voltage to the effective concentrations of a cell reaction's constituent parts. The total of the components for migration, diffusion, and convection may be used to indicate the flow of ions (referred to as i) via ion exchanger. The Nernst-Einstein equation provides a definition for species movement in the bed.

$$\bar{u}_i = \overline{D}_i z_i F / RT$$

where, $\bar{u}_i$ is the mobility of ion in ion exchanger (m^2/Vs), $\overline{D}_i$ is the diffusion coefficient (m^2/s)T.

2.5.7 Transport mechanism in ion-exchange resins

An electrical potential and electroactive materials are used in the EDI process to extract ionic species from solutions. In EDI devices, the electroactive medium may operate to constantly move ions via ionic or electronic replacement processes, or it could alternatively collect and release ionizable species. Devices for EDI may use media with a fixed or transient charge and could be used batch-wise, constantly, or periodically. The resin improves the movement of ions and serves as a platform for electrochemical processes such as hydrogen and hydroxyl ionization of water. There are several medium designs that may be used, including closely mixed AER and CER and distinct portions of IER that are largely made up of the same polarization of resin, such as AER/CER (Wood et al., 2010).

The two unique functioning regimes for EDI devices are electroregeneration and improved transfer. The resins inside the unit stay in their salt forms during the increased transfer regime. Ions are transported through the chambers to the surface of the IEM using the IER, which is the orders of magnitude greater conducting than the fluid in low conductivity situations. In order to preserve electrical neutrality, this method of ion removal can only be used in systems that permit the elimination of both anions and cations at the same time. The continual renewal of resins by both H^+ and OH^- from the electrostatically water splitting is what distinguishes the electro regeneration regime. The ion-depleting compartment's bipolar interfaces are where this dissociation happens most frequently (Arar et al., 2011).

2.6 Conclusion

EDI has immense promise in a variety of fields, such as the creation of distilled water as well as selective extraction and concentrating. Due to the absence of a thorough grasp of its principles, progress was relatively sluggish. For the ongoing evolution of this technique, knowing the precise kinetics of these systems is critical. Outstanding ionic extraction efficiency, reduced process expenditures, and residue-free operation are the main advantages of this environmentally friendly, greener separation method. Electrostatic insulation, enhanced novel materials, nanoparticles, and wafer ion exchange have the potential to dramatically increase the performance of EDI systems as novel electrode coatings with higher activity and larger surface areas evolve. EDI systems are projected to continue to advance because they operate more cost-effectively and efficiently because of their minimal energy use makes them desirable for commercial production in several worldwide applications.

References

Al-Amshawee, S., Yunus, M.Y.B.M., Azoddein, A.A.M., Hassell, D.G., Dakhil, I.H., Hasan, H.A., 2020. Electrodialysis desalination for water and wastewater: a review. Chemical Engineering Journal 380, 122231.

Alvarado, L., Chen, A., 2014. Electrodeionization: principles, strategies and applications. Electrochimica Acta 132, 583−597.

Alvarado, L., Torres, I.R., Chen, A., 2013. Integration of ion exchange and electrodeionization as a new approach for the continuous treatment of hexavalent chromium wastewater. Separation and Purification Technology 105, 55−62.

Arar, Ö., Yüksel, Ü., Kabay, N., Yüksel, M., 2014. Various applications of electrodeionization (EDI) method for water treatment—a short review. Desalination 342, 16−22.

Arar, Ö., Yüksel, Ü., Kabay, N., Yüksel, M., 2011. Removal of Cu^{2+} ions by a microflow electrodeionization (EDI) system. Desalination 277 (1−3), 296−300.

Bazinet, L.F.L.L., Lamarche, F., Ippersiel, D., 1998. Bipolar-membrane electrodialysis: applications of electrodialysis in the food industry. Trends in Food Science & Technology 9 (3), 107−114.

Berrios, M., Siles, J.A., Martín, M.A., Martín, A., 2013. Ion exchange. Separation and Purification Technologies in Biorefineries. Wiley, pp. 149−165.

Campione, A., Gurreri, L., Ciofalo, M., Micale, G., Tamburini, A., Cipollina, A., 2018. Electrodialysis for water desalination: a critical assessment of recent developments on process fundamentals, models and applications. Desalination 434, 121−160.

Chen, T.L., Chen, L.H., Lin, Y.J., Yu, C.P., Ma, H.W., Chiang, P.C., 2021. Advanced ammonia nitrogen removal and recovery technology using electrokinetic and stripping process towards a sustainable nitrogen cycle: a review. Journal of Cleaner Production 309, 127369.

Dammak, L., Fouilloux, J., Bdiri, M., Larchet, C., Renard, E., Baklouti, L., et al., 2021. A review on ion-exchange membrane fouling during the electrodialysis process in the food industry, part 1: types, effects, characterization methods, fouling mechanisms and interactions. Membranes 11 (10), 789.

Ebbers, B., Ottosen, L.M., Jensen, P.E., 2015. Electrodialytic treatment of municipal wastewater and sludge for the removal of heavy metals and recovery of phosphorus. Electrochimica Acta 181, 90−99.

Ervan, Y., Wenten, I.G., 2002. Study on the influence of applied voltage and feed concentration on the performance of electrodeionization. Songklanakarin Journal of Science and Technology 24, 955−963.

Esmaeili, H.O.S.S.E.I.N., Foroutan, R.A.U.F., 2015. Investigation into ion exchange and adsorption methods for removing heavy metals from aqueous solutions. International Journal of Biology, Pharmacy and Allied Sciences 4, 620−629.

Feng, X., Gao, J.S., Wu, Z.C., 2008. Removal of copper ions from electroplating rinse water using electrodeionization. Journal of Zhejiang University-SCIENCE A 9 (9), 1283−1287.

Gifford, J.D., Atnoor, D.E.V.E.N., 2000, October. An innovative approach to continuous electrodeionization module and system design for power applications. In International Water Conference (pp. 22−26). Pittsburg, PA. Oetober.

Gu, J., Liu, H., Wang, S., Zhang, M., Liu, Y., 2019. An innovative anaerobic MBR–reverse osmosis-ion exchange process for energy-efficient reclamation of municipal wastewater to NEWater-like product water. Journal of Cleaner Production 230, 1287−1293.

Hakim, A.N., Khoiruddin, K., Ariono, D., Wenten, I.G., 2020. Ionic separation in electrodeionization system: mass transfer mechanism and factor affecting separation performance. Separation & Purification Reviews 49 (4), 294−316.

Ijanu, E.M., Kamaruddin, M.A., Norashiddin, F.A., 2020. Coffee processing wastewater treatment: a critical review on current treatment technologies with a proposed alternative. Applied Water Science 10, 1−11.

Jaroszek, H., Dydo, P., 2016. Ion-exchange membranes in chemical synthesis—a review. Open Chemistry 14 (1), 1−19.

Jina, Q., Yaob, W., Chenc, X., 2020. Removal of Cr (VI) from wastewater by simplified electrodeionization. Desalination And Water Treatment 183, 301−306.

Jordan, M.L., Valentino, L., Nazyrynbekova, N., Palakkal, V.M., Kole, S., Bhattacharya, D., et al., 2020. Promoting water-splitting in Janus bipolar ion-exchange resin wafers for electrodeionization. Molecular Systems Design & Engineering 5 (5), 922−935.

Juve, J.M.A., Christensen, F.M., Wang, Y., Wei, Z., 2022. Electrodialysis for metal removal and recovery: a review. Chemical Engineering Journal 134857.

Kammerer, J., Carle, R., Kammerer, D.R., 2011. Adsorption and ion exchange: basic principles and their application in food processing. Journal of agricultural and food chemistry 59 (1), 22−42.

Kansara, N., Bhati, L., Narang, M., Vaishnavi, R., 2016. Wastewater treatment by ion exchange method: a review of past and recent researches. ESAIJ (Environmental Science, An Indian Journal) 12 (4), 143−150.

Khoiruddin, K., Hakim, A.N., Wenten, I.G., 2014. Advances in electrodeionization technology for ionic separation—a review. Membrane and Water Treatment 5 (2), 87−108.

Kumar, S., Jain, S., 2013. History, introduction, and kinetics of ion exchange materials. Journal of Chemistry 2013.

Kumar, P.S., Varsha, M., Rathi, B.S., Rangasamy, G., 2022. Electrodeionization: fundamentals, methods and applications. Environmental Research 114756.

Lee, H., Jin, Y., Hong, S., 2016. Recent transitions in ultrapure water (UPW) technology: rising role of reverse osmosis (RO). Desalination 399, 185−197.

Lee, D., Lee, J.Y., Kim, Y., Moon, S.H., 2017. Investigation of the performance determinants in the treatment of arsenic-contaminated water by continuous electrodeionization. Separation and Purification Technology 179, 381−392.

Liu, Y., Deng, Y.Y., Zhang, Q., Liu, H., 2021. Overview of recent developments of resource recovery from wastewater via electrochemistry-based technologies. Science of The Total Environment 757, 143901.

Lu, J., Ma, X.Y., Wang, Y.X., 2016. Numerical simulation of the electrodeionization (EDI) process with layered resin bed for deeply separating salt ions. Desalination and Water Treatment 57 (23), 10546−10559.

Lu, J., Wang, Y.X., Lu, Y.Y., Wang, G.L., Kong, L., Zhu, J., 2010. Numerical simulation of the electrodeionization (EDI) process for producing ultrapure water. Electrochimica Acta 55 (24), 7188−7198.

Lu, H., Wang, Y., Wang, J., 2014. Removal and recovery of Ni2 + from electroplating rinse water using electrodeionization reversal. Desalination 348, 74−81.

Lu, H.X., Wang, J.Y., Bu, S.F., Zhang, M.Q., Zhang, J.B., 2011. Removal of nickel ions from dilute heavy metal solutions by electrodeionization process, Advanced Materials Research, 183. Trans Tech Publications Ltd, pp. 580−584.

Ma, A., Abushaikha, A., Allen, S.J., McKay, G., 2019. Ion exchange homogeneous surface diffusion modelling by binary site resin for the removal of nickel ions from wastewater in fixed beds. Chemical Engineering Journal 358, 1−10.

Meaden, G.T., 2013. Electrical Resistance of Metals. Springer.

Mikhaylin, S., Bazinet, L., 2016. Fouling on ion-exchange membranes: classification, characterization and strategies of prevention and control. Advances in Colloid and Interface Science 229, 34−56.

Mir, N., Bicer, Y., 2021. Integration of electrodialysis with renewable energy sources for sustainable freshwater production: a review. Journal of Environmental Management 289, 112496.

Mistry, G., Popat, K., Patel, J., Panchal, K., Ngo, H.H., Bilal, M., et al., 2022. New outlook on hazardous pollutants in the wastewater environment: occurrence, risk assessment and elimination by electrodeionization technologies. Environmental Research 219, 115112.

Mohammadi, R., Tang, W., Sillanpää, M., 2021. A systematic review and statistical analysis of nutrient recovery from municipal wastewater by electrodialysis. Desalination 498, 114626.

Moon, S.H., Yun, S.H., 2014. Process integration of electrodialysis for a cleaner environment. Current Opinion in Chemical Engineering 4, 25−31.

Muhammad, A., Soares, A., Jefferson, B., 2019. The impact of background wastewater constituents on the selectivity and capacity of a hybrid ion exchange resin for phosphorus removal from wastewater. Chemosphere 224, 494−501.

Narayanan, S., Marucheck, A.S., Handfield, R.B., 2009. Electronic data interchange: research review and future directions. Decision Sciences 40 (1), 121−163.

Nasef, M.M., Ujang, Z., 2012. Introduction to ion exchange processes. Ion Exchange Technology I: Theory and Materials. Springer, pp. 1−39.

Nazif, A., Karkhanechi, H., Saljoughi, E., Mousavi, S.M., Matsuyama, H., 2022. Recent progress in membrane development, affecting parameters, and applications of reverse electrodialysis: a review. Journal of Water Process Engineering 47, 102706.

Nemati, M., Hosseini, S.M., Shabanian, M., 2017. Novel electrodialysis cation exchange membrane prepared by 2-acrylamido-2-methylpropane sulfonic acid; heavy metal ions removal. Journal of Hazardous Materials 337, 90−104.

Otero, C., Urbina, A., Rivero, E.P., Rodriguez, F.A., 2021. Desalination of brackish water by electrodeionization: experimental study and mathematical modeling. Desalination 504, 114803.

Ottosen, L.M., Kristensen, I.V., Pedersen, A.J., Hansen, H.K., Villumsen, A., Ribeiro, A. B., 2003. Electrodialytic removal of heavy metals from different solid waste products. Separation Science and Technology 38 (6), 1269−1289.

Pang, H., He, J., Ma, Y., Pan, X., Zheng, Y., Yu, H., et al., 2021. Enhancing volatile fatty acids production from waste activated sludge by a novel cation-exchange resin assistant strategy. Journal of Cleaner Production 278, 123236.

Pedersen, A.J., Ottosen, L.M., Villumsen, A., 2005. Electrodialytic removal of heavy metals from municipal solid waste incineration fly ash using ammonium citrate as assisting agent. Journal of Hazardous Materials 122 (1–2), 103–109.

Rathi, B.S., Kumar, P.S., Parthiban, R., 2022. A review on recent advances in electrodeionization for various environmental applications. Chemosphere 289, 133223.

Rathi, B.S., Kumar, P.S., 2020. Electrodeionization theory, mechanism and environmental applications. A review. Environmental Chemistry Letters 18, 1209–1227.

Sarıçiçek, E.N., Tuğaç, M.M., Özdemir, V.T., İpek, İ.Y., Arar, Ö., 2021. Removal of boron by boron selective resin-filled electrodeionization. Environmental Technology & Innovation 23, 101742.

Shen, X., Liu, Q., Li, H., Kuang, X., 2022. Membrane-free electrodeionization using graphene composite electrode to purify copper-containing wastewater. Water Science & Technology 86 (7), 1733–1744.

Sivakumar, D., Shankar, D., Gomathi, V., Nandakumaar, A., 2014. Application of electro-dialysis on removal of heavy metals. Pollution Research 33, 627–637.

Song, J.H., Song, M.C., Yeon, K.H., Kim, J.B., Lee, K.J., Moon, S.H., 2005. Purification of a primary coolant in a nuclear power plant using a magnetic filter—electrodeionization hybrid separation system. Journal of Radioanalytical and Nuclear Chemistry 262, 725–732.

Strathmann, H., 2010a. Electrodialysis, a mature technology with a multitude of new applications. Desalination 264 (3), 268–288.

Strathmann, H., 2010b. Ion-exchange membrane processes in water treatment. Sustainability Science and Engineering 2, 141–199.

Strathmann, H., Krol, J.J., Rapp, H.J., Eigenberger, G., 1997. Limiting current density and water dissociation in bipolar membranes. Journal of Membrane Science 125 (1), 123–142.

Sun, L., Chen, Q., Lu, H., Wang, J., Zhao, J., Li, P., 2020. Electrodialysis with porous membrane for bioproduct separation: technology, features, and progress. Food Research International 137, 109343.

Sun, T.R., Ottosen, L.M., 2012. Effects of pulse current on energy consumption and removal of heavy metals during electrodialytic soil remediation. Electrochimica Acta 86, 28–35.

Tanaka, Y., 2007. Limiting current density. Membrane Science and Technology 12, 245–270.

Tanaka, Y., 2005. Limiting current density of an ion-exchange membrane and of an electrodialyzer. Journal of Membrane Science 266 (1–2), 6–17.

Tate, J., 2014. Comparison of continuous electrodeionization technologies. IWC 14, 01.

Tian, H., Wang, Y., Pei, Y., Crittenden, J.C., 2020. Unique applications and improvements of reverse electrodialysis: a review and outlook. Applied Energy 262, 114482.

Ulusoy Erol, H.B., Hestekin, C.N., Hestekin, J.A., 2021. Effects of resin chemistries on the selective removal of industrially relevant metal ions using wafer-enhanced electrodeionization. Membranes 11 (1), 45.

Valero, F., Barceló, A., Arbós, R., 2011. Electrodialysis technology: theory and applications. Desalination, Trends and Technologies 28, 3–20.

Wang, J., Wan, Z., 2015. Treatment and disposal of spent radioactive ion-exchange resins produced in the nuclear industry. Progress in Nuclear Energy 78, 47–55.

Wenten, I.G., Khoiruddin, K., Alkhadra, M.A., Tian, H., Bazant, M.Z., 2020. Novel ionic separation mechanisms in electrically driven membrane processes. Advances in Colloid and Interface Science 284, 102269.

Widiasa, I.N., Sutrisna, P.D., Wenten, I.G., 2004. Performance of a novel electrodeionization technique during citric acid recovery. Separation and Purification Technology 39 (1–2), 89–97.

Wolfaardt, F.J., Fernandes, L.G.L., Oliveira, S.K.C., Duret, X., Görgens, J.F., Lavoie, J. M., 2021. Recovery approaches for sulfuric acid from the concentrated acid hydrolysis of lignocellulosic feedstocks: a mini-review. Energy Conversion and Management: X 10, 100074.

Wood, J., Gifford, J., Arba, J., Shaw, M., 2010. Production of ultrapure water by continuous electrodeionization. Desalination 250 (3), 973–976.

Xing, Y., Chen, X., Wang, D., 2009. Variable effects on the performance of continuous electrodeionization for the removal of Cr (VI) from wastewater. Separation and Purification Technology 68 (3), 357–362.

Yeon, K.H., Song, J.H., Moon, S.H., 2004. A study on stack configuration of continuous electrodeionization for removal of heavy metal ions from the primary coolant of a nuclear power plant. Water research 38 (7), 1911–1921.

Zahakifar, F., Keshtkar, A.R., Souderjani, E.Z., Moosavian, M.A., 2020. Use of response surface methodology for optimization of thorium (IV) removal from aqueous solutions by electrodeionization (EDI). Progress in Nuclear Energy 124, 103335.

CHAPTER THREE

Configuration and mechanism of electrodeionization module

3.1 Introduction

Electrodeionization (EDI) can generate exceptionally pure water. This technique is preferable in several ways, such as the absence of chemicals, environmental friendliness, and ease of use (Su et al., 2013). Because of the existence of organic compounds that serve as the junction between the ion-exchange membranes (IEMs), dilute chambers behave as conductors when filled with ion-exchange resin (IER). This method proved effective in combating electrodialysis (ED) concentration polarization, as indicated by a 50% improvement in highest ion removal efficiency to 90%. As a result, the beneficial combination of ED and ion exchange (IE), which incorporates the advantages of both techniques, successfully overcomes some operational difficulties related to each method separately (Alvarado and Chen, 2014).

The electrodes are separated by cationic exchange membrane (CEM) and anionic exchange membranes (AEMs), much like in an ED device (Tseng et al., 2022). The middle chamber is loaded with IER that improve cationic or anionic transport while being driven by direct current (DC). Hydrogen (H^+) and hydroxyl (OH^-) ions are produced by the water breakdown mechanism during the EDI process (Rathi et al., 2022). Without the need of chemical reagents, the electrochemical reaction-generated H^+ and OH^- ions constantly replenish the IER. Alternate AEMs and CEM are common components of EDI devices. Liquid flow chambers with inlets and outputs are created by configuring the gaps between the membranes (Jordan et al., 2023). Electrodes are positioned at the endpoints of the membranes and chambers to apply a transverse DC electric current. The ion-depleting chambers of a continuous EDI device are loaded with IER. The IER improves ion movement and can function as a substrate in electrochemical processes like the electrochemical breakdown of water into hydrogen ion and hydroxyl ions (Qian et al., 2022).

Electrodeionization: Fundamentals, Methods and Applications.
DOI: https://doi.org/10.1016/B978-0-443-18983-8.00003-X
© 2024 Elsevier Inc. All rights reserved, including those for text and data mining, AI training, and similar technologies.

Several media arrangements are feasible in an EDI device, such as segregated IER sections or intimately mixed anion exchange resin (AER) and cation exchange resins (CER). Each part was mostly made up of identical polarity IER, such as AER, CER, or mixed resins (Arar et al., 2014).

Very pure water is produced through EDI, together with the removal of ions from contaminants and the absence of chemical reagents needed for resin renewal. The disadvantages of resin beds for IE are solved by EDI, notably the discharge of ions when the beds expire (Zhou et al., 2022). Applications include the elimination of radioactive contaminants, heavy metal ions, toxic chemicals, hardness, nitrates, ammonia, and the generation of deionized water and desalination (Rathi and Kumar, 2020). This chapter investigates the EDI technique for ion ejection and transport, which comprises anions, cations, and both anion and cation elimination. It also looked at the transport methodologies provided by the IERs utilized in EDI, such as AER, CER, and mixed bed resin. Adsorption and desorption concepts that impact mass transfer in EDI were also addressed.

3.2 Electrodeionization mechanism for expulsion and movement of ions

An EDI device is made up of electrode sections, concentrated chamber, and diluted chamber that are loaded with heterogeneous IER. With heterogeneous resins acting as a conductive path, electrical field effect causes anions and cations that are in solution to be directed to the anode and cathode, accordingly. The electrolysis of water reaction generates a significant amount of hydrogen ion and hydroxyl ions, which will continually replenish the mixed IER. Hence without the need for chemicals' renewal, EDI may be continually run. Utilizing IER as the bridging over flow is a technique to reduce the impedance and the energy usage of the ED stack since dilute solution has a comparatively large impedance. This approach can be also employed to effectively treat dilute aqueous solutions (Widiasa and Wenten, 2014).

3.2.1 Anion expulsion

Anions are negatively charged due to having one extra electron or many (Tate, 2014). AER with positively charged groups, such alkyl substituted phosphine, amino groups, alkyl substituted sulfides, and others, are attached to the materials' backbones and permit anions to travel but refuse cations (Kumar and Jain, 2013). Table 3.1 represents the different anion removal with its process condition and removal percentage. In the ecosystem, arsenic is a naturally existing metalloid that is exceedingly mobility. Its movement is greatly influenced by the source elemental form, the level of redox, and the mobilization processes. Arsenite and arsenate are found in both reducing and oxidized settings due to delayed redox reactions. All biological forms are identified as highly poisonous to arsenic. The primary factors for removing heavy metals are the chemistry and constitution of arsenic-contaminated water. Since that arsenite is primarily noncharged at pH levels under 9.2, the majority of extraction methods are more effective for arsenate. As a result, there are less opportunities for adsorption, IE, or precipitation with the trivalent form of arsenic (Nicomel et al., 2016).

Lee et al. (2017)'s study evaluated the As (V) removal from wastewater using EDI. To test the cell saturation threshold, voltage from 0.8 to 5.2 V by 0.2 V step input was administered with a 3-hour retention time. Impedance, ionic strength, and pH of each chamber of the continous electrodeionization (CEDI) unit were examined in conjunction at each voltage procedure to generate an ion movement and ion-exchange framework. Hydrogen ion and hydroxide ions are present in the IER at first. As the process went on, the hydrogen and the hydroxide ions in the IER were exchanged by the arsenate and sodium ions, and those ions were again depleted into the flowing fluid by an electrical current and a concentration gradient. The released hydroxide ions and hydrogen ion traveled in opposite directions, toward the positive and negative electrodes (Rathi and Kumar, 2022). The anolyte become acidic and the catholyte turned basic as an outcome. A specific spot had a pH flip as the voltage raised. The resin-containing compartment's recirculating vessel's dilute chamber experiences ionic strength degradation at various initial concentrations. Arsenic is eliminated more quickly at 5 mg/L initial concentration than at 15 mg/L at 15 V cell voltage. This tendency is explained by the simplicity of removing arsenic at lower initial concentrations when there are sufficient hydroxide ions to support the exchange balance at the AER (Ortega et al., 2017).

The ability of a proposed EDI technique to eradicate nitrate from water under the influence of various species was tested in a number of bench scale studies (Meyer et al., 2005b; Yeon and Moon, 2001).

Table 3.1 Different anion removal by electrodeionization with its process condition and removal percentage.

Anion	Initial concentration (ppm)	Flow rate	pH	Time	Energy consumption	Removal percentage (%)	References
Chromium (Anionic bed)	100	12 mL/min	5	1.3 h	0.91 kW h/m^3	97.55	Alvarado et al. (2009)
Chromium (Mixed bed)	100	12 mL/min	5	1.3 h	0.167 kWh/m^3	99.8	Alvarado et al. (2009)
Chromium (Mixed bed)	100	4.5 mL/min.	3	4 h	—	99	Zhang et al. (2014)
Chromium (Anionic bed)	100	4.5 mL/min	6.88−7.21	8 h	0.07 c	98.5	Alvarado et al. (2013)
Chromium (Anionic bed)	50	1.2 L/h	—	2 h	8.4 kWh/mol	95.5	Jina et al. (2020)
Chromium (Anionic bed)	100	2.0 L/h	8	6 h	7.3 kWh/mol	99.5	Xing et al. (2009)
Chromium (Cationic bed)	25	2.0 L/h	2.75	2 h	—	98.5	Arar et al. (2011)
Arsenic (Anionic bed)	15 ppm	20 mL/min	—	12 h	60.7 kW h/kg	98.8	Rathi and Kumar (2022)
Arsenic (Mixed Bed)	10	1 mL/min	10	12 h	1.5 kWh/m^3	99.5	Lee et al. (2017)
Arsenic (Anionic bed)	15	30 mL/min	7.3	8.5 h	7.5 kW h/kg	99.5	Ortega et al. (2017)
Nitrate (Mixed Bed)	200	12 mL/min	7	4 h	—	96.25	Zhang and Chen (2016)
Nitrate (Mixed Bed)	300	12 mL/min	7	10 h	1.7 Wh/L	99	Bi et al. (2011)

Pure water that had been ingested with certain anionic and cationic species as well as two freshwaters with two different dissolved solids contents were used in the studies. In order to determine whether the technique could maintain nitrate eradication over a prolonged period of time, longer duration trials were conducted. It was discovered that using IER in the EDI unit led to a procedure which was greater energy- and nitrate-selective than when using a broad AER (Meyer et al., 2005a). The EDI with mixed bed of various AER and CER (2:1) was successful in removing nitrate intensity of wastewater sample from diluted compartment that had increased quickly in a relatively short time after passing electric current and could be depleted to less than 10 ppm and satisfactory the benchmark of the WHO. In the process of regenerating IER, electric current was utilized in place of reagents, and the resins' renewal efficacy reached 60% after 90 hours (Bi et al., 2011).

A VoltaLab was used to apply the limited power to the unit for the purpose of removing nitrate from both the real groundwater and the synthesized wastewater. Then, 1.0 L of nitrate solution was circulated within the exact same tank as the feed water by pumping it into the diluted chamber at predetermined flow conditions. In the meantime, 250 mL of a 0.1 M sulfuric acid solution was continuously redistributed into the electrodes' rinse chambers. Nitrate ions moved to the anode when the cell received current as a result of the electrical field's presence. The concentration of nitrate ions risen in the anionic chamber despite the semipermeable membranes that separated the distinct sections. In the concentrated sections, cations and anions increased in population while being removed from the diluted section at the end of the operation. To measure the amounts of nitrate ions, specimens from the diluted and concentrated chambers were periodically removed during batch-mode EDI studies (Zhang and Chen, 2016).

An EDI technique is used to create a device for the continuous operation of fluoride-free freshwater. With the aid of a research laboratory device, the tests are conducted. The EDI method usually runs in two schemes: electroregeneration and improved transfer. Water breakup caused the second phase of the EDI system's current performance to decline. At the beginning, deionized water was delivered into the concentrated chamber while groundwater with a specified amount and level of fluoride solution was supplied into the dilute chamber. In a bid to preserve the agitation, peristaltic pumps were employed to feed the solutions of 1000 cm^3 into the corresponding chambers in a recirculating mode at a steady rate of flow 0.01 m³/h. Constant voltage was applied across the electrodes using a DC power source, and the ensuing current fluctuation was observed as a function of period. Throughout all of the trials, the conductance and pH electrodes were placed in the appropriate vessels to continuously measure variations in the conductance and pH at the outlet from the dilute compartment (DLC) and the concentrate compartment (CC). After 15 minutes, the fluoride content in the permeate was thoroughly inspected. The method also has the advantage of continuously removing species to an incredibly significant level while being both economical and environmentally friendly (Gahlot et al., 2015). Fig. 3.1 depicts the ion transport in anion expulsion using AER in EDI.

Using an extended cell technique, in which the mixed resins were put into the diluted section, Cr(III) and Cr(VI) were separated and recovered. The feed stream to the diluted chamber had a flow velocity of 4.5 mL/min

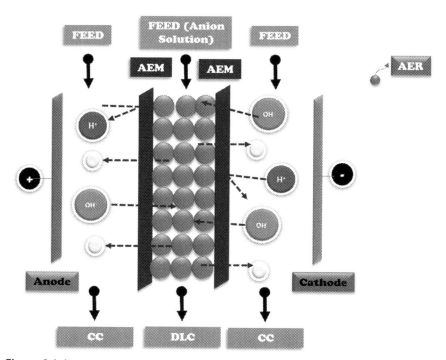

Figure 3.1 Ion transport in anion expulsion using AER in EDI. *AEM*, Anion exchange membrane; *AER*, anion exchange resin; *CC*, concentrate compartment; *DLC*, dilute compartment; *EDI*, electrodeionization.

and included a combination of Cr(III) and Cr(VI) with starting concentrations of 100 mg/L each. In order to prevent deposition due to the process's decrease of pH, the concentrated chambers were supplied concurrently with 0.1 M H_2SO_4. In the electrode washing chamber, 200 mL of a 0.1 M sulfuric acid solution was further cycled (Zhang et al., 2014). The Cr(VI) solution was fed into the dilute chamber, and the potassium hydroxide solution was pumped to the washing chamber. Potassium hydroxide served as the supplementary electrolyte that carried the electrical current in each chamber. An 'immediate oversupply of power' may be produced at the interface among anionic and cationic species due to a brief drop in the ions whenever the cell system was exposed to an electric current in the diluted chamber. The breaking of the water might be brought on by the immediate oversupply of power (Alvarado et al., 2013). At pH 5, the first chromium ions are Cr(III) and Cr (VI), which are present in the resin at positions 1 and 2, correspondingly. The ion percentage will alter during the ion-exchange procedure as a consequence of the pH rise brought on by the replacement of Cr(VI) by hydroxyl ion at pH = 7.97 (Alvarado et al., 2009).

3.2.2 Cation expulsion

Sulfate, benzoate, phosphate, carboxylate, and other negatively charged group-containing cation materials are attached to the framework element and let cations to enter but rejecting anions (Kumar and Jain, 2013). Table 3.2 represents the different cation removal with its process condition and removal percentage.

A brand-new method of ED and EDI does not require a membrane to remove copper ions from industrial wastewaters. The suggested novel EDI method varies from traditional ED, continuous EDI methods in that it does not employ any semipermeable IEM, and as a result, it does not display constraints related to membranes, such as concentration polarization and water breakage (Dermentzis et al., 2009). The feed was added to the diluted chambers as a solution containing copper sulfate. In the concentrated chambers, a new copper sulfate solution was cycled. Some persisted in the electrode wash solution, while part of these copper ions deposition on the cathode. Both the quantity of copper added to the concentrated chambers and the amount taken out of the diluted streams are copper. A rather large initial current was attained once the voltage was introduced. The resistance quickly rose as ions were withdrawn from the diluted sections, and the current declined sharply to its lowest value.

Table 3.2 Different Cation removal by electrodeionization with its process condition and removal percentage.

Cation	Initial concentration	Flow rate	pH	Time	Removal percentage (%)	References
Copper (Mixed bed)	100 ppm	0.129 mL/s	2.5	10 h	99	Dermentzis et al. (2009)
Copper (Mixed bed)	50 ppm	1.24 L/h	2.35	8 h	99.9	Guan and Wang (2007)
Copper (Mixed bed)	50 ppm	3 L/h	4.28	8 h	99.5	Feng et al. (2008)
Copper (Cation bed)	100 ppm	30 L/h	3.2	10 h	99	Mahmoud et al. (2007)
Nickel (Mixed bed)	87.9 ppm	3 L/h	2.5	12 h	99.9	Feng et al. (2007)
Copper (Mixed bed)	94.9 ppm	3 L/h	2.5	12 h	99.9	Feng et al. (2007)
Zinc (Mixed bed)	63.4 ppm	3 L/h	2.5	12 h	99.9	Feng et al. (2007)
Cadmium (Mixed bed)	98.1 ppm	3 L/h	2.5	12 h	99.9	Feng et al. (2007)
Nickel (Electrostatically Shielded)	100 ppm	$2.02 \times 10^{-4} \text{ dm}^3/\text{s}$	4	10 h	99.9	Dermentzis (2010a,b)
Nickel (Cation bed)	300 mol/m^3	$1.67 \times 10^{-5} \text{ m}^3/\text{s}$	2.8	8 h	99.9	Dzyazko (2006)
Nickel (Cation bed)	602 mol/m^3	$2 \times 10^{-7} \text{ m}^3/\text{s}$	4	8 h	99	Spoor et al. (2001b)
Nickel (Mixed bed)	50 ppm	2.5 L/h	5.7	8 h	99.8	Lu et al. (2010)
Nickel (Mixed bed)	50 ppm	10 L/h	3	10 h	97	Lu et al. (2014)
Nickel (Mixed bed)	1000 ppm	—	—	3 h	99.8	Wardani et al. (2017b)
Nickel (Mixed bed)	300 ppm	—	—	1 h	99	Wardani et al. (2017b)
Nickel (Cation bed)	0.3 mol/m^3	$1.2 \text{ dm}^3/\text{h}$	3	20 h	99.9	Dzyazko et al. (2014)
Nickel (Mixed bed)	50 ppm	25 L/h	5	10 h	99.8	Lu et al. (2015)
Nickel (Mixed bed)	50 ppm	6 L/h	5.7	7 h	99	Lu et al. (2011b)
Nickel (Cation bed)	82 mol/m^3	$1.7 \times 10^{-5} \text{ m}^3/\text{s}$	3.5	12 h	99.9	Rozhdestvenska et al. (2006)

Nickel (Mixed bed)	55 ppm	5 L/h	6.27	8 h	99	Lu et al. (2011a)
Cobalt (Mixed bed)	10 ppm	100 mL/min	5	1.5 h	97	Yeon et al. (2003)
Cobalt (Mixed bed)	0.34–mM	5 mL/min	5	2 h	99	Yeon et al. (2004)
Cobalt (Electrostatically Shielded)	300 ppm	4.06×10^{-4} dm^3/s	3	40 min	84.36	Dermentzis et al. (2010b)
Cobalt (Mixed bed)	5000 ng/L	30 L/h	3	7.27 h	99	Li et al. (2010)

Then, when ions collected in the concentrated chambers, the stack impedance decreased and the current gradually increased (Guan and Wang, 2007). The enhancement and elimination of copper ions benefited from high supply voltage. A reduction in current efficiency resulted from the additional hydrogen and hydroxyl ions that the increased potential induced to migrate to the concentrated chamber from the electrode processes. Hence, for greater enhancement, better elimination, and less energy usage, a suitable applied voltage ought to be employed (Feng et al., 2007). Fig. 3.2 depicts the ion transport in cation expulsion using CER in EDI.

In Spoor et al. (2001b), it was investigated how to remove nickel ions from a low cross-linked IER utilizing an applied electrical voltage difference. As the substance had to be partly packed in order to participate in an EDI process, the nickel dosage was not nearly at the ion exchanger limit. Insoluble deposits of nickel hydroxide complex cannot accumulate

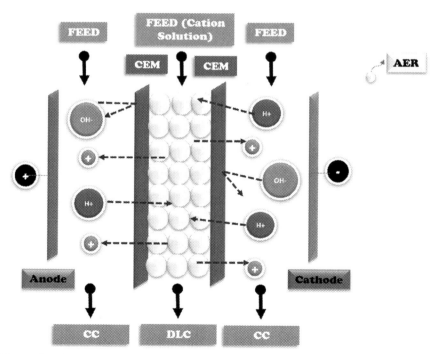

Figure 3.2 Ion transport in cation expulsion using CER in EDI. *CC*, Concentrate compartment; *CEM*, cation exchange membrane; *CER*, cation exchange resin; *DLC*, dilute compartment; *EDI*, electrodeionization.

on the outside of the IER/IEM since hydrogen ion can balance out hydroxide ions that form there. Nickel cations moved to the cathode chamber during the course of the study. Although the cathode chamber was flooded with acid that had a rather high concentration, no nickel formation was seen. In this instance, the cathode only produced hydrogen (Dzyazko, 2006; Spoor et al., 2001a). In order to create a chamber in the cell known as the cathodic protection chamber, a new CEM was introduced between the cathode and the last AEM next to it. The cathodic protection chamber may be capable of minimizing nickel deoxidization on the cathode and stop the movement of hydroxide ions generated by the cathode reaction from the cathode to the CC. As a result, there was significantly less nickel hydroxide deposition. To create an anodic protected chamber, an extra AEM was put between the CEM next to the anode and the AEM of the first DLC. The oxidation damage to AEM of the first DLC caused by anode reaction may then be avoided, as well as the cationic transfer from the anodic chamber to the first DLC (Lu et al., 2010).

To eliminate cobalt ions, an EDI procedure was developed. The cobalt ion is removed using two different sorts of processes. The first is ion movement that is compelled by the applied electric field, while the other refers to the interchange of cobalt ions with protons produced by the bipolar membrane within the IER. Once an electrical potential was supplied to the stack, more than 97% removal efficiency was attained in just 10 minutes. The cell potential varied somewhat, most likely as a result of alterations to electrical impedance (Yeon et al., 2003). The CEDI system was run with the CER bed ahead of the MER bed, the AER bed, and the multilayered bed. Three sections made up the IER bed: the bottom section, which had CER to remove metal; the middle section, which held AER to remove anions; and the top section, which included a mixed bed to regulate pH. This stack arrangement accomplished over 99% ion elimination by preventing an interaction between the metal ions and hydroxide ions (Yeon et al., 2004). The use of a DC electric field can prevent breakthrough and increase removal effectiveness. As a result, cobalt elimination and desalination happen simultaneously and constantly without building up in the resin substance. Since there would theoretically be no wasted resin formation if CEDI stack was used in primary coolant processing in nuclear power plant, it has the advantage over the conventional ion-exchange technique of minimizing nuclear waste (Li et al., 2010).

3.2.3 Anion and cation expulsion

The charge distinguishes anions from cations. A positively charged ion is referred to as a cation when it has additional protons compared to electrons. An ion is called an anion when it has a net negative charge due to the presence of more electrons than protons. Table 3.3 represents simultaneous anion and cation removal by EDI with its process condition and energy consumption

A beneficial method for desalinating waters with salinities in the range of saltwater is EDI. Although modeling can aid in understanding how EDI cells behave, no trustworthy models integrating important performance factors have yet to be developed. The AEM and CEM of the DLC have a different electric potential in EDI. The fluid is in motion as it passes across the interstitial gap created by the mixed bed AER and CER. Two possible distributions occur in the resin bed as a result of CER and AER's ability to carry electrical currents, or cations and anions, accordingly. Because of the forces generated by migrations, dispersion, and convection processes, the ions in the liquid state are moved toward or away from the resins or liquid surfaces (Otero et al., 2021).

IE or EDI are two methods for producing high purity water. In many ways, the later deionization process is preferable to the earlier, notably the need for no chemicals, low environmental impact, and simplicity of use. The methods by which DC power may cause an ion-exchange reaction's equilibrium to alter in a way that was advantageous for the renewal of resins were complicated. One of them ought to be connected to the water breakup that often takes place when resin particles come into touch with one another. Saline ions may be drawn off one curved surface of a resin particle and into solution when subjected to a high electric field. The ion-exchange equilibrium was shifted at the same time as water split on the opposite curved surface of this resin particle, supplying hydrogen ion or hydroxyl ions that could penetrate into the resin particle to replace the migration ions. The alteration in the attractive forces interactions between the resins and the ions adsorbed may have been another significant process that led to the shift in the ion-exchange equilibrium. The attractive electrostatic interactions were likely decreased because of the applied current, which led to the desorption of the absorbed ions until a new adsorption−desorption equilibrium was restored (Su et al., 2013). It is commonly known that cation ions travel toward a cathode in an electric field, whereas anion ions move toward an anode. The effectiveness of a

Table 3.3 Simultaneous anion and cation removal by electrodeionization with its process condition and energy consumption.

Application	Initial concentration of salt/hardness	Flow rate	Applied voltage	Time	Energy consumption	References
Desalination	4000 ppm	5 L/h	19 V	2 h	3.71 kWh/m^3	Sun et al. (2016)
Desalination	3000 ppm	5 L/h	13 V	2 h	1.95 kWh/m^3	Sun et al. (2016)
Desalination	2000 ppm	5 L/h	11 V	2 h	1.04 kWh/m^3	Sun et al. (2016)
Desalination	5000 ppm	1.1 L/min	3.45 V	2 h	0.657 kWh/m^3	Pan et al. (2017)
Desalination	3000 ppm	0.81 L/min	2.28 V	1.5 h	0.66 kWh/m^3	Zheng et al. (2018)
Desalination	0.5 M	10 m/h	1400 v	1 h	1.5 kWh/m^3	Shen et al. (2014)
Desalination	7.9 ppm	32 L/h	8 V	150 h	0.03 kWh/m^3	Liu et al. (2019)
High purity water production	26 ppm	15 m/h	720 V	60 h	0.68 kWh/m^3	Su et al. (2013)
High purity water production	10 ppm	1.0 m^3/h	30 V	400 h	—	Wenten and Arfianto (2013)
High purity water production	100 ppm	14 L/h	848 V	5 h	0.71 kWh/m^3	Su et al. (2014)
High purity water production	2.82 ppm	15 m/h	407 V	48 h	0.41 kWh/m^3	Hu et al. (2015)
High purity water production	11.7 ppm	30 L/h	30 V	130 h	0.26 kWh/m^3	Su et al. (2010)
High purity water production	5 ppm	15 m/h	433 V	72 h	0.24 kWh/m^3	Hu et al. (2016)
High purity water production	6.1 ppm	0.5 m/s	50 V	3 h	0.89 kWh/m^3	Wardani et al. (2017a)

(*Continued*)

Table 3.3 (Continued)

Application	Initial concentration of salt/hardness	Flow rate	Applied voltage	Time	Energy consumption	References
High purity water production	35 ppm	7.5 L/h	25 V	6 h	0.658 kWh/kg	Bhadja et al. (2015)
Water softening	250 ppm	200 mL/min	20 V	3 h	—	Lee et al. (2012)
Water softening	40.5 ppm	60 mL/min	20 V	0.5 h	24.2 Wh/L	Lee et al. (2013)
Water softening	5 mg/dm^3	26 dm^3/h	17.5 V	9 h	—	Fu et al. (2009)
Water softening	300 ppm	15 L/h	11.3 V	200 h	23.6 kWh/kg	Jin et al. (2019)
Water softening	350 ppm	40 L/h	—	12 h	3.17 kWh/kg	Yu et al. (2018)

thin AER layer in preventing cation ion-reversed migration in membrane-free EDI was examined by contrasting resin electrical renewal with and without a thin layer of strong-base resin. When exposed to an electric potential, the concentration of sodium ions and calcium ions in the top cation resin with and without a thin layer of strong-base resin has been monitored. Whereas the calcium ion concentration indicated the reverse emigration scenario of the calcium ions, the sodium ion concentration in the upper strong-acid cation resin indicated the reverse emigration scenario of the sodium ions. The sodium ion and calcium ion concentrations were 54 mg/L resin and 5.1 mg/L resin, correspondingly, when the thin strong-base resin layer was not employed, demonstrating a clear backward movement of cation ions during renewal. Sodium ion concentration was almost ten times higher than calcium ion concentration. In other words, monovalent ions passed through resin particles more easily than divalent ions (Su et al., 2014). Fig. 3.3 depicts the ion transport in anion and cation expulsion using AER and CER in EDI.

Deep softening from synthetic water solution with low concentrations of total hardness was accomplished using an uniquely designed EDI stack

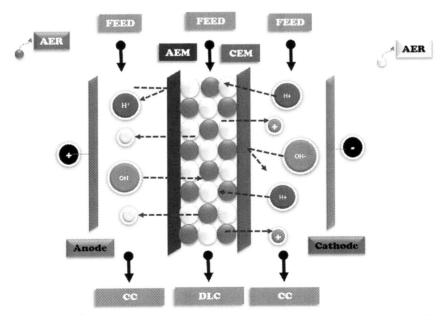

Figure 3.3 Ion transport in anion and cation expulsion using AER and CER in EDI. *AER*, Anion exchange resin; *CER*, cation exchange resin; *EDI*, electrodeionization.

(Fu et al., 2009). There were primarily two steps in the calcium ion removal procedure. A thin calcium carbonate layer that acted as the crystal's nuclei was produced in the first stage as a result of reactions that took place on the cathode's surface. The precipitate would develop more quickly as the crystalline nucleus grew. This is what caused the first improvement in calcium ion removal efficiency. Later, while precipitation reactions continued, the process progressively moved on to the following step. Precipitation had mostly coated the cathode surface at this point. The effective sedimentation area shrunk as a result. In addition, as the precipitates collected, the reaction zone moved farther from the cathode surface and the ion transit pathways among the precipitates concurrently diminished. The calcium removal percentage therefore decreased as a consequence of a decrease in the convection and diffusive velocities of calcium ions and carbonate ions. Magnesium ions was in a unique circumstance, and its percentage removal did not decrease with time. This one was due to the mass transfer constraint having less of an impact on the precipitate of magnesium than it did on calcium ions, when HCO_3^- initially interacted with hydroxide ions. Moreover, Mg^{2+} has a lower ionic radius than Ca^{2+}. This was in the diffusion's advantage (Jin et al., 2019).

3.3 Transport mechanisms created by the ion-exchange resins used in electrodeionization

IER is essentially used to modify the ED process to create EDI. EDI is a membrane-based technique used to separate, to put it another way. Ionized substances are eliminated in EDI in a manner similar to that of traditional ED, with the ion percentage removal being significantly accelerated by the inclusion of IER in the DLC. Others believe that improved transfer and electroregeneration are the two separate working domains for EDI devices. The resins in the apparatus continue to exist as salt develops in the improved transfer region. On the resin surfaces in this region, IE serves as the transportation route. Electroregeneration is a term used to describe the second operational region for EDI equipment. Resins are continuously renewed in this region by electrically powered water-splitting processes. The resins are regenerated into their hydroxides and hydrogen states (Meng et al., 2004).

IE in EDI involves two steps of procedures in the chambers that IER fills. Ions move from the bulk liquid electrolyte to the IER surface in the first step. After then, moving ions on the IER interchange opposite charge ions in the IER interface. The bound counter ions in the IER are moved by voltage difference to the IEM in the second step. After passing via IEM, the ions ultimately accumulate in the CC (Hakim et al., 2020). A straightforward AER or CER bed partitioned by two membranes, wherein no deionization processes of the solution occur only a swap of any of the solution ions for the corresponding ion of the resin. This is especially helpful when one of the ions has to be removed but the existence of the resin's ion in the final solution has no negative effects on it and could possibly be desired (Ortega et al., 2017).

In a DLC, the contact forms of IER and IEM may be categorized into four different groups. AER contacts with AEM and CER in Type A, and CER interactions with AER and CEM, correspondingly. IER and IEM in this scenario allow ions that exchange between the AER and CER to pass easily from the DLC to the CC, which causes an immediate decrease in the level of ions between the resins and subsequent splitting of the water. According to Type B, the CER makes connection with the AEM and AER, and the AER makes connection with the CER and CEM, respectively. In this instance, the water–splitting phenomenon and the phenomenon of quickly falling ion concentration will occur at the associated contacting surfaces of AEM and CER. Type C demonstrates that the two AERs make touch with one another before making connection with the corresponding AEM and CEM. AER can facilitate the rapid movement of anion ions from the DLC to the CC, and in this situation, splitting of water will take place at the AER-CEM interface surface. On the other hand, when two CER resins come into contact with one another as well as with AEM and CEM, accordingly, water–splitting will occur at the AEM-CER contacting face, and CER resins can rapidly transport cation ions from the DLC to the CC (Meng et al., 2004).

In DLC, the anion exchange process between the liquid and IER occurs heterogeneously. It is considered that the interchange of arsenic ions is evident of the ion-exchange mechanism in DLC even though the IE among arsenic anions and hydroxide relies on the predominant arsenic species. As a result, the ion-exchange procedure can be split into two stages. (1) The transfer of species via the diffusion layer that envelops resin particle. (2) Ion-exchange reaction and ion movement in the IER from or toward the interface. Ionic particles are disseminated and carried by the liquid stream while this process takes place and the liquid passes through the packed bed (Rivero et al., 2018).

In the EDI procedure, ions are basically fully transported by the IER and are not impacted by water impedance. The mixed ion-exchange resin (MER) serves in this procedure as a conductive medium. When the DLC's free ions are insufficient to support current transmission across the medium, a water-splitting reaction takes place in those DLC, and comparatively high concentrations of hydrogen ions and hydroxide ions are then capable of continually renewing the MER. A MER column with continuous regeneration can be compared to the EDI unit's distinctive 'electro-regeneration', which makes it able to perform entire deionization process (Widiasa et al., 2004).

The DLC is often filled with MER, which removes ions from solutions and causes the wastewater to become somewhat deionized. Its principal drawbacks are its poor current efficacy and limited capacity for deionizing mixtures of weak acids and bases. Another arrangement involves inserting individual AER and CER in the first bed, where any of the solution ions, such as sodium, is swapped for the resin ion, H^+, keeping the number of ions roughly the same. The ion elimination or EDI is really accomplished in the subsequent bed because the secondary ion of the solution, such as Cl^-, is exchanged for the resin ion, OH^-, as it flows via the secondary bed. A third arrangement entails employing separate AER and CER beds with a bipolar membrane among the two beds. The regeneration of resins is made possible by the hydrogen ions and hydroxide ions produced by water splitting at the bipolar membranes (Ortega et al., 2017). In EDI, the resin filler's capability for exchange of ion is far less significant than its ability to quickly transfer ions to the IEM's surface. As a result, the IERs are not optimized for capacity but rather for other factors that affect transfer, such selectivity and retention of water (Wood et al., 2010).

3.4 Principles of adsorption/desorption that affect mass transport in electrodeionization

A lot of research on mass transfer in the EDI procedure have been published. A summary of the research, IER plays an important function in enabling ionic transport. The IER keeps the chamber's conductivity extremely high in order to keep the ionic flow high. A number of investigations have indicated that IERs with a high concentration of functional groups, resulting in a high ion-exchange capacity, good conductivity, and small particle size distribution, are desirable. Isolation of ions with identical charges is also feasible, which is related to the IER's attraction and preferential transport. Nevertheless, the role of IER in the manufacturing of high purity water or a low-dose solution is only being researched. The IER has a greater conductivity of ions than the treated solution in this instance. The mechanism of water dissociation is a crucial phenomenon in the EDI mechanism. It often occurs at IER—IER or IER—IEM interfaces. Water dissociation improves EDI performance by converting IER to hydrogen or hydroxide forms, which have greater conductivity values than their salt counterparts. This transition enables EDI to undertake deep deionization, which is very useful in the manufacture of ultrapure water. Water dissociation, on the other hand, affects total current efficacy and therefore be optimized (Hakim et al., 2020).

Spoor et al. (2002) discovered that nickel ion movement was affected by the voltage difference, the size of the IER bed, the initial level of ion adsorbed in CER, and the electrolyte level in the anode chamber. Furthermore, the ionic flow is proportional to the applied voltage difference. Migrated ions from CER are replenished by hydrogen ions from the anolyte solution. In CER, both ionic flux and current efficacy grow exponentially, whereas ionic mobility declines with the concentration of ions. Ions migrated primarily via the IER when its electric conductivity is greater than that of the treated solution. As a result, IER with high conductance is recommended. If the IER is entirely loaded with ions, electromigration or no ionic adsorption dominates the process of elimination rate. In addition, due to the differences in attraction and preferential transfer, IER may aid in the separation of ions with an identical electrical charge.

More research has been conducted on the utilization of amine-based IER as CO_2 adsorbents via thermal swing. Those resins are said to be stable after many adsorption—desorption cycles, with only a slight decline in capture capability, and to need fairly mild desorption parameters (Sharifian et al., 2021).

In the literature, several fundamental EDI arrangements have been examined, including (1) distinct beds, (2) beds divided by a bipolar membrane, (3) simple beds (AER or CER), and (4) mixed beds (MER), as seen in Fig. 3.4 The last method is often used to produce high purity

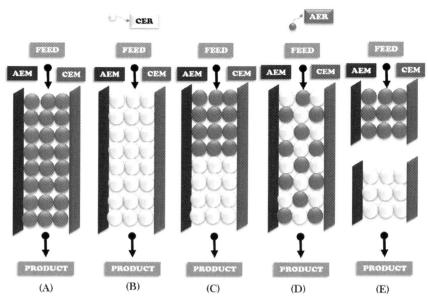

Figure 3.4 Ion-exchange resin arrangements in EDI (A) anion exchange bed EDI, (B) cation exchange bed EDI, (C) layered bed EDI, (D) mixed bed EDI, and (E) separate bed EDI. *EDI*, Electrodeionization.

water from ion-deficient solutions. Additionally, the MER bed enhances mass transfer, makes resin renewal easier, and decreases desalted solution impedance. Literature research has revealed that the process is mass transfer regulated with in terms of the movement of ions in the DLC and IE at the IER/liquid interface. They employed a bed of CER between two CEM for EDI of copper solutions in a simple bed instead of an AER bed. A mass transfer coefficient was used to describe the ion flow in n-CSTR that were utilized for modeling EDI (Otero et al., 2021).

3.5 Conclusion

This chapter examines the EDI method of ion transport and ejection, which involves the removal of both cations and anions. Additionally, it examined the transport strategies offered by the IERs used in EDI, including mixed bed resin, AER, and CER. Also, Adsorption and desorption ideas that affect EDI mass transfer were covered. Operating factors including the voltage difference, electrical current density, flow velocity, and quality of feed have an impact on how well EDI performs. Important criteria for module design include things such as cell thickness, chamber length, stack arrangement, IER type and shape, and IEM. It has been established that EDI is an excellent, environmentally acceptable method for concentrating, isolating, and purifying. In order to advance and expand this technology, new compounds are continually being developed, which might have significant positive effects on the global economy and ecological research.

References

Alvarado, L., Chen, A., 2014. Electrodeionization: principles, strategies and applications. Electrochimica Acta 132, 583−597.

Alvarado, L., Ramírez, A., Rodríguez-Torres, I., 2009. Cr(VI) removal by continuous electrodeionization: study of its basic technologies. Desalination 249 (1), 423−428.

Alvarado, L., Torres, I.R., Chen, A., 2013. Integration of ion exchange and electrodeionization as a new approach for the continuous treatment of hexavalent chromium wastewater. Separation and Purification Technology 105, 55−62.

Arar, Ö., Yüksel, Ü., Kabay, N., Yüksel, M., 2011. Removal of Cu^{2+} ions by a microflow electrodeionization (EDI) system. Desalination 277 (1−3), 296−300.

Arar, Ö., Yüksel, Ü., Kabay, N., Yüksel, M., 2014. Various applications of electrodeionization (EDI) method for water treatment—a short review. Desalination 342, 16−22.

Bhadja, V., Makwana, B.S., Maiti, S., Sharma, S., Chatterjee, U., 2015. Comparative efficacy study of different types of ion exchange membranes for production of ultrapure water via electrodeionization. Industrial & Engineering Chemistry Research 54 (44), 10974−10982.

Bi, J., Peng, C., Xu, H., Ahmed, A.S., 2011. Removal of nitrate from groundwater using the technology of electrodialysis and electrodeionization. Desalination and Water Treatment 34 (1−3), 394−401.

Dermentzis, K., 2010a. Removal of nickel from electroplating rinse waters using electrostatic shielding electrodialysis/electrodeionization. Journal of Hazardous Materials 173 (1−3), 647−652.

Dermentzis, K., Davidis, A., Papadopoulou, D., Christoforidis, A., Ouzounis, K., 2009. Copper removal from industrial wastewaters by means of electrostatic shielding driven electrodeionization. Journal of Engineering Science & Technology Review 2 (1).

Dermentzis, K.I., Davidis, A.E., Dermentzi, A.S., Chatzichristou, C.D., 2010b. An electrostatic shielding-based coupled electrodialysis/electrodeionization process for removal of cobalt ions from aqueous solutions. Water Science and Technology 62 (8), 1947−1953.

Dzyazko, Y.S., 2006. Purification of a diluted solution containing nickel using electrodeionization. Desalination 198 (1−3), 47−55.

Dzyazko, Y.S., Ponomaryova, L.N., Rozhdestvenskaya, L.M., Vasilyuk, S.L., Belyakov, V.N., 2014. Electrodeionization of low-concentrated multicomponent Ni^{2+}-containing solutions using organic−inorganic ion-exchanger. Desalination 342, 43−51.

Feng, X., Wu, Z., Chen, X., 2007. Removal of metal ions from electroplating effluent by EDI process and recycle of purified water. Separation and purification Technology 57 (2), 257−263.

Feng, X., Gao, J.S., Wu, Z.C., 2008. Removal of copper ions from electroplating rinse water using electrodeionization. Journal of Zhejiang University-SCIENCE A 9 (9), 1283−1287.

Fu, L., Wang, J., Su, Y., 2009. Removal of low concentrations of hardness ions from aqueous solutions using electrodeionization process. Separation and Purification Technology 68 (3), 390−396.

Gahlot, S., Sharma, S., Kulshrestha, V., 2015. Electrodeionization: an efficient way for removal of fluoride from tap water using an aluminum form of phosphomethylated resin. Industrial & Engineering Chemistry Research 54 (16), 4664−4671.

Guan, S., Wang, S., 2007. Experimental studies on electrodeionization for the removal of copper ions from dilute solutions. Separation Science and Technology 42 (5), 949−961.

Hakim, A.N., Khoiruddin, K., Ariono, D., Wenten, I.G., 2020. Ionic separation in electrodeionization system: mass transfer mechanism and factor affecting separation performance. Separation & Purification Reviews 49 (4), 294−316.

Hu, J., Fang, Z., Jiang, X., Li, T., Chen, X., 2015. Membrane-free electrodeionization using strong-type resins for high purity water production. Separation and Purification Technology 144, 90−96.

Hu, J., Chen, Y., Zhu, L., Qian, Z., Chen, X., 2016. Production of high purity water using membrane-free electrodeionization with improved resin layer structure. Separation And Purification Technology 164, 89−96.

Jin, H., Yu, Y., Zhang, L., Yan, R., Chen, X., 2019. Polarity reversal electrochemical process for water softening. Separation and Purification Technology 210, 943−949.

Jina, Q., Yaob, W., Chenc, X., 2020. Removal of Cr (VI) from wastewater by simplified electrodeionization. Desalination And Water Treatment 183, 301−306.

Jordan, M.L., Kokoszka, G., Gallage Dona, H.K., Senadheera, D.I., Kumar, R., Lin, Y.J. et al., 2023. Integrated Ion-Exchange Membrane Resin Wafer Assemblies for Aromatic Organic Acid Separations Using Electrodeionization. ACS Sustainable Chemistry & Engineering.

Kumar, S., Jain, S., 2013. History, introduction, and kinetics of ion exchange materials. Journal of Chemistry 2013.

Lee, H.J., Hong, M.K., Moon, S.H., 2012. A feasibility study on water softening by electrodeionization with the periodic polarity change. Desalination 284, 221−227.

Lee, H.J., Song, J.H., Moon, S.H., 2013. Comparison of electrodialysis reversal (EDR) and electrodeionization reversal (EDIR) for water softening. Desalination 314, 43−49.

Lee, D., Lee, J.Y., Kim, Y., Moon, S.H., 2017. Investigation of the performance determinants in the treatment of arsenic-contaminated water by continuous electrodeionization. Separation and Purification Technology 179, 381−392.

Li, F.Z., Zhang, M., Zhao, X., Hou, T., Liu, L.J., 2010. Removal of Co^{2+} and Sr^{2+} from a primary coolant by continuous electrodeionization packed with weak base anion exchange resin. Nuclear Technology 172 (1), 71−76.

Liu, Y., Wang, J., Xu, Y., Wu, B., 2019. A deep desalination and anti-scaling electrodeionization (EDI) process for high purity water preparation. Desalination 468, 114075.

Lu, H., Wang, J., Yan, B., Bu, S., 2010. Recovery of nickel ions from simulated electroplating rinse water by electrodeionization process. Water Science and Technology 61 (3), 729−735.

Lu, H.X., Wang, J.Y., Bu, S.F., Zhang, M.Q., Zhang, J.B., 2011a. Removal of nickel ions from dilute heavy metal solutions by electrodeionization process, Advanced Materials Research, 183. Trans Tech Publications Ltd, pp. 580−584.

Lu, H., Wang, J., Bu, S., Fu, L., 2011b. Influence of resin particle size distribution on the performance of electrodeionization process for Ni^{2+} removal from synthetic wastewater. Separation Science and Technology 46 (3), 404−408.

Lu, H., Wang, Y., Wang, J., 2014. Removal and recovery of Ni^{2+} from electroplating rinse water using electrodeionization reversal. Desalination 348, 74−81.

Lu, H., Wang, Y., Wang, J., 2015. Recovery of Ni^{2+} and pure water from electroplating rinse wastewater by an integrated two-stage electrodeionization process. Journal of Cleaner Production 92, 257−266.

Mahmoud, A., Muhr, L., Grévillot, G., Lapicque, F., 2007. Experimental tests and modelling of an electrodeionization cell for the treatment of dilute copper solutions. The Canadian Journal of Chemical Engineering 85 (2), 171−179.

Meng, H., Peng, C., Song, S., Deng, D., 2004. Electro-regeneration mechanism of ion-exchange resins in electrodeionization. Surface Review and Letters 11 (06), 599−605.

Meyer, N., Parker, W.J., Van Geel, P.J., Adiga, M., 2005a. Development of an electrodeionization process for removal of nitrate from drinking water Part 2: Multi-species testing. Desalination 175 (2), 167−177.

Meyer, N., Parker, W.J., Van Geel, P.J., Adiga, M., 2005b. Development of an electrodeionization process for removal of nitrate from drinking water Part 1: Single-species testing. Desalination 175 (2), 153−165.

Nicomel, N.R., Leus, K., Folens, K., Van Der Voort, P., Du Laing, G., 2016. Technologies for arsenic removal from water: current status and future perspectives. International Journal of Environmental Research and Public Health 13 (1), 62.

Ortega, A., Oliva, I., Contreras, K.E., González, I., Cruz-Díaz, M.R., Rivero, E.P., 2017. Arsenic removal from water by hybrid electro-regenerated anion exchange resin/electrodialysis process. Separation and Purification Technology 184, 319−326.

Otero, C., Urbina, A., Rivero, E.P., Rodriguez, F.A., 2021. Desalination of brackish water by electrodeionization: experimental study and mathematical modeling. Desalination 504, 114803.

Pan, S.Y., Snyder, S.W., Ma, H.W., Lin, Y.J., Chiang, P.C., 2017. Development of a resin wafer electrodeionization process for impaired water desalination with high energy efficiency and productivity. Acs Sustainable Chemistry & Engineering 5 (4), 2942−2948.

Qian, F., Lu, J., Gu, D., Li, G., Liu, Y., Rao, P., et al., 2022. Modeling and optimization of electrodeionization process for the energy-saving of ultrapure water production. Journal of Cleaner Production 372, 133754.

Rathi, B.S., Kumar, P.S., 2020. Electrodeionization theory, mechanism and environmental applications. A review. Environmental Chemistry Letters 18, 1209−1227.

Rathi, B.S., Kumar, P.S., 2022. Continuous electrodeionization on the removal of toxic pollutant from aqueous solution. Chemosphere 291, 132808.

Rathi, B.S., Kumar, P.S., Parthiban, R., 2022. A review on recent advances in electrodeionization for various environmental applications. Chemosphere 289, 133223.

Rivero, E.P., Ortega, A., Cruz-Díaz, M.R., González, I., 2018. Modelling the transport of ions and electrochemical regeneration of the resin in a hybrid ion exchange/electrodialysis process for As (V) removal. Journal of Applied Electrochemistry 48, 597−610.

Rozhdestvenska, L.M., Dzyazko, Y.S., Belyakov, V.N., 2006. Electrodeionization of a Ni^{2+} solution using highly hydrated zirconium hydrophosphate. Desalination 198 (1−3), 247−255.

Sharifian, R., Wagterveld, R.M., Digdaya, I.A., Xiang, C., Vermaas, D.A., 2021. Electrochemical carbon dioxide capture to close the carbon cycle. Energy & Environmental Science 14 (2), 781−814.

Shen, X., Li, T., Jiang, X., Chen, X., 2014. Desalination of water with high conductivity using membrane-free electrodeionization. Separation and Purification Technology 128, 39−44.

Spoor, P.B., Veen, W.T., Janssen, L.J.J., 2001a. Electrodeionization 1: migration of nickel ions absorbed in a rigid, macroporous cation-exchange resin. Journal of Applied Electrochemistry 31, 523–530.

Spoor, P.B., Ter Veen, W.R., Janssen, L.J.J., 2001b. Electrodeionization 2: the migration of nickel ions absorbed in a flexible ion-exchange resin. Journal of Applied Electrochemistry 31, 1071–1077.

Spoor, P.B., Koene, L., Ter Veen, W.R., Janssen, L.J.J., 2002. Electrodeionisation 3: the removal of nickel ions from dilute solutions. Journal of applied electrochemistry 32, 1–10.

Su, Y., Wang, J., Fu, L., 2010. Pure water production from aqueous solution containing low concentration hardness ions by electrodeionization. Desalination and Water Treatment 22 (1–3), 9–16.

Su, W., Pan, R., Xiao, Y., Chen, X., 2013. Membrane-free electrodeionization for high purity water production. Desalination 329, 86–92.

Su, W., Li, T., Jiang, X., Chen, X., 2014. Membrane-free electrodeionization without electrode polarity reversal for high purity water production. Desalination 345, 50–55.

Sun, X., Lu, H., Wang, J., 2016. Brackish water desalination using electrodeionization reversal. Chemical Engineering and Processing: Process Intensification 104, 262–270.

Tate, J., 2014. Comparison of continuous electrodeionization technologies. IWC 14, 01.

Tseng, P.C., Lin, Z.Z., Chen, T.L., Lin, Y., Chiang, P.C., 2022. Performance evaluation of resin wafer electrodeionization for cooling tower blowdown water reclamation. Sustainable Environment Research 32 (1), 36.

Wardani, A.K., Hakim, A.N., Wenten, I.G., 2017a. Combined ultrafiltration-electrodeionization technique for production of high purity water. Water Science and Technology 75 (12), 2891–2899.

Wardani, A.K., Hakim, A.N., Khoiruddin, Destifen, W., Goenawan, A., Wenten, I.G., 2017b, January. Study on the influence of applied voltage and feed concentration on the performance of electrodeionization in nickel recovery from electroplating wastewater. In AIP Conference Proceedings (Vol. 1805, No. 1, p. 030004). AIP Publishing LLC.

Wenten, I.G., Arfianto, F., 2013. Bench scale electrodeionization for high pressure boiler feed water. Desalination 314, 109–114.

Widiasa, I.N., Wenten, I.G., 2014. Removal of inorganic contaminants in sugar refining process using electrodeionization. Journal of Food Engineering 133, 40–45.

Widiasa, I.N., Sutrisna, P.D., Wenten, I.G., 2004. Performance of a novel electrodeionization technique during citric acid recovery. Separation and Purification Technology 39 (1–2), 89–97.

Wood, J., Gifford, J., Arba, J., Shaw, M., 2010. Production of ultrapure water by continuous electrodeionization. Desalination 250 (3), 973–976.

Xing, Y., Chen, X., Yao, P., Wang, D., 2009. Continuous electrodeionization for removal and recovery of Cr (VI) from wastewater. Separation and Purification Technology 67 (2), 123–126.

Yeon, S.H., Moon, K.H., 2001. Principle and application of continuous electrodeionization. Membrane Journal 11 (2), 61–65.

Yeon, K.H., Seong, J.H., Rengaraj, S., Moon, S.H., 2003. Electrochemical characterization of ion-exchange resin beds and removal of cobalt by electrodeionization for high purity water production. Separation Science and Technology 38 (2), 443–462.

Yeon, K.H., Song, J.H., Moon, S.H., 2004. A study on stack configuration of continuous electrodeionization for removal of heavy metal ions from the primary coolant of a nuclear power plant. Water Research 38 (7), 1911–1921.

Yu, Y., Jin, H., Quan, X., Hong, B., Chen, X., 2018. Continuous multistage electrochemical precipitation reactor for water softening. Industrial & Engineering Chemistry Research 58 (1), 461–468.

Zhang, Z., Chen, A., 2016. Simultaneous removal of nitrate and hardness ions from groundwater using electrodeionization. Separation and Purification Technology 164, 107−113.

Zhang, Z., Liba, D., Alvarado, L., Chen, A., 2014. Separation and recovery of Cr (III) and Cr (VI) using electrodeionization as an efficient approach. Separation and Purification Technology 137, 86−93.

Zheng, X.Y., Pan, S.Y., Tseng, P.C., Zheng, H.L., Chiang, P.C,, 2018. Optimization of resin wafer electrodeionization for brackish water desalination. Separation and Purification Technology 194, 346−354.

Zhou, X., Yan, G., Majdi, H.S., Le, B.N., Khadimallah, M.A., Ali, H.E., et al., 2022. Spotlighting of microbial electrodeionization cells for sustainable wastewater treatment: Application of machine learning. Environmental Research 115113.

CHAPTER FOUR

Construction of electrodeionization

4.1 Introduction

The procedure of electrodeionization (EDI), which integrates electrodialysis (ED) with ion exchange (IE), has been used for producing ultrapure water (UPW). The primary factor in the business's profitability is the elimination of chemical regeneration (Hakim and Wenten, 2014). EDI is a green procedure, which is probably the main factor in its commercial viability. By replacing the dangerous chemicals often employed to renew IE resins (IERs) with electrical power, it gets rid of the waste flow connected to IER renewal. Semipermeable IE membranes (IEMs), together with continuously loaded media such as IER, are commonly found in EDI equipment (Rathi et al., 2022).

Alternate permselective cation exchange membranes (CEMs) and anion exchange membranes (AEMs) are common components of EDI devices. The membrane gaps are designed to form compartments for liquid flow with inlets and outputs (Nikonenko et al., 2018). The electrodes at both ends of the IEM and compartments are used in conjunction with a separate power source to apply an inverse DC electrical field. Ions in the fluid are drawn to their corresponding opposite electrodes when the chambers are exposed to an electrical current. Due to the ion depletion that occurs, the chambers delimited by the AEM toward the anode and the CEM toward the cathode are known as diluting compartments (DLCs). The concentrating compartments (CCs), which are defined as those delimited by the AEM approaching the cathode and the CEM approaching the anode, will then "capture" ions. In an EDI equipment, IER is placed inside of the DLC or both the DLC and the CC. In addition to improving ion transportation, the IER can function as a substrate in electrochemical processes (ECRs), such as the electrically charged dissociation of water into hydrogen and hydroxyl ions (Wood et al., 2010).

IEMs have changed over the past 50 years from being a laboratory tool to industrial goods with substantial technological and economic significance. IEMs are effectively used nowadays to treat industrial wastewater and desalinate sea and saltwater. They are gaining a lot of attention. They are effective instruments for concentrating or removal of ionic species-containing foods, pharmaceuticals, and basic chemical goods. The creation of an IEM makes societies environmentally friendly by recovering usable effluents that would otherwise go to waste products, in addition to making the procedure more environmentally friendly and energy-efficient (Xu, 2005). Synthetic IER is increasingly being used to research natural media. These resources provide novel and creative solutions to the issues that conventional methods are unable to address. However, there seems to be uncertainty over the correct application of IER for soil and environmental investigations due to a lack of expertise or limited expertise (Skogley and Dobermann, 1996). Anode and cathode are coupled to a source of continuous electric current, which maintains IER activation while an electrical field among the anode and cathode promotes the electrolysis of water and the removal of the dissolved ions. Anion exchange, cation exchange, and mixed bed EDI are a few examples of the EDI types based on how the major parts of the EDI unit are arranged. The main goal of this chapter was to examine the electrodes and various kinds of IER, IEM, and electrodes that make up the EDI unit. Additionally, the numerous types of EDI and the procedure's chemistry were covered.

4.2 Overview of electrodeionization

In an EDI stack or unit, permselective AEM and CEM are often placed in alternately succession between two electrodes. A separator is placed between both sets of AEM and CEM to create a chamber. An EDI stack has three distinct chamber types such as DLC, CC, and EC. IER is put inside each compartment (Hu et al., 2016; Lu et al., 2010a,b,c). The IER improves ionic species transportation by raising the total stack conductance from the bulk solution to the IEM. In a plate-and-frame module, typical industrial EDI stacks are found. There are other stack types available, including spiral wrapped and stacked-disk stacks (Wood, 2008). Anions and cations move in the direction of the anode and cathode, correspondingly, when a solution of electrolytes is introduced into EDI chambers and an electrical field is generated at the points of the electrodes. The cations are rejected by the AEM but flow via the CEM. According to Wood and Gifford (2003), anions also travel via AEM and are rejected by CEM (Gifford and Atnoor, 2000).

While it is focused on the CC, the electrolyte concentration of the solution is decreased in the DLC. A solution with little concentration or substantial resistance might be effectively addressed with IER (Hu et al., 2015). When it comes to the manufacturing of HPW, the IER also makes splitting of water possible, making it possible to effectively remove species that are weakly ionized, including boron and silica. Water is divided into hydrogen and hydroxide ions by the DC electrical field, and this in return continually regenerates the IER (Shen et al., 2019). The system gets cleaned out with the swapped ions once they have been moved via the IEM to the CC. The method of EDI could be utilized for either concentrating or purifying since it generates two separate streams with varying ion concentrations. By executing a thorough separation in the DLC or by achieving a high enriching factor in the CC, it is possible to EDI can either make HPW or retrieve valuable goods (Hakim et al., 2020).

4.2.1 Ion-exchange resin

IERs are insoluble polymers with an ion-active side chains and a polystyrene (PS) backbone that has been cross-linked (Bresser et al., 2018). The IER is frequently utilized to reduce the disagreeable tastes of medications and stop them from sticking to the tongues and oral cavity surfaces. Several uses for the treatment of water employ IER. Water softening and demineralization of water are among the most important of these uses in terms of the quantity of IER required. By switching out the positively charged particles for sodium ions and maybe replacing the anions for sodium chloride ions, IERs serve to soften water. According to Bajpai (2018/2018), they can additionally be utilized to demineralize water by substituting hydrogen ions for the cations and hydroxyl ions for the anions. According to Vagliasindi et al. (1998), there are two basic forms of IER: cation-exchange resin (CER) and anion-exchange resin (AER). CER exchanges positively charged particles such as sodium for calcium while AER exchanges negative-charged ions such as chloride for arsenic. Table 4.1 gives the IER employed in different application of EDI.

In rare instances, both the DLC and the CC of an EDI device's DLC have been loaded with IER. The ECR, which splits water into ions of hydrogen

Table 4.1 Ion-exchange resin employed in electrodeionization.

Application	Anion exchange resin (AER)	Cation exchange resin (CER)	References
Arsenic removal	Amberlite IRA-400	Amberlite IR 120	Lee et al. (2017)
Boron removal	Purolite A 500	Purolite CT 175	Arar et al. (2013)
Calcium removal	Amberlite IRA-400 chloride form	Amberlite IR 120	Kurup et al. (2009)
Cesium removal	—	Amberlite IR 120	Zahakifar et al. (2017)
Chromium removal	Amberlite IRA900RF Cl	Amberlite 200C Na	Alvarado et al. (2013)
Chromium removal	Amberlite IRA-67	Dowex Mac-3	Alvarado et al. (2009)
Chromium removal	IRA201 × 4	—	Xing et al. (2009b)
Chromium removal	IRA 312	—	Xing et al. (2009b)
Chromium removal	IRA D201	—	Xing et al. (2009b)
Chromium removal	IRA D354	—	Xing et al. (2009b)
Chromium removal	Amberlite IRA-67	Dowex Mac-3	Alvarado et al. (2015)
Chromium removal	IRA201 × 4	—	Xing et al. (2009a)
Chromium removal	IRA201 × 4	—	Jina et al. (2020)
Chromium removal	IRA 312	—	Jina et al. (2020)
Chromium removal	IRA D301	—	Jina et al. (2020)
Chromium removal	IRA D201	—	Jina et al. (2020)
Cobalt removal	201 × 7 strong base styrene	001 × 7 strong acid styrene	Li et al. (2010)
Cobalt removal	Purolite A847 gel-type weak base	001 × 7 strong acid styrene	Li et al. (2010)
Cobalt removal	Polyurethane IRN78	Polyurethane IRN77	Yeon and Moon (2003)
Cobalt removal	Amberlite IRN78	Amberlite IRN77	Yeon et al. (2004)
Copper Removal	Styrene-base AER D296	Styrene -base CER D072	Guan and Wang (2007)
Copper Removal	—	Purolite C-150PLH	Arar et al. (2011)
Hardness Removal	Amberlite 200C Na	Amberlite IRA900RF Cl	Zhang and Chen (2016)

Hardness removal	201 × 7 strong base styrene	001 × 7 strong acid styrene	Fu et al. (2009)
Lithium removal	Purolite A500 plus	Purolite C145	Demir et al. (2023)
Nitrate removal	Amberlite 200C Na	Amberlite IRA900RF Cl	Zhang and Chen (2016)
Nitrate removal	201 × 7 strong base styrene	001 × 7 strong acid styrene	Bi et al. (2011)
Nitrate removal	Macroporous D407	Macroporous D001	Bi et al. (2011)
Nitrate removal	Amberlite IRN78	Amberlite IRN77	Yeon et al. (2004)
Production of pure water	—	Amberlite IRN 77	Yeon et al. (2003)
Production of pure water	—	Diaion SKN 1	Yeon et al. (2003)
Production of pure water	—	Amberlite IR 120	Yeon et al. (2003)
Purification of organic acids	Amberlite IRA-400 chloride form	Amberlite IR 120	Lopez and Hestekin (2015)
Recovery of nickel	Styrene-base AER D296	Styrene-base CER D072	Lu et al. (2015)
Recovery of nickel	Styrene-base anion resins D296	Styrene -base cation resins D072	Lu et al. (2010a,b,c)
Recovery of nickel	—	Dowex MSC-1	Dzyazko (2006)
Silica removal	Purolite A 500	Purolite CT 175	Arar et al. (2013)
Strontium removal	—	Amberlite IR 120	Zahakifar et al. (2017)
Sugar refining process	Strong base type I AER, A-400	Purolite strong acid CER, C-100E	Widiasa et al. (2014)
Thorium removal	Amberlite IRA-400	Amberlite IR 120	Zahakifar et al. (2020)

and hydroxyl, can take place with the help of the IER, which also improves the movement of ions (Sarıçiçek et al., 2021). Different media arrangements, including closely mixed bed IER (MIER) or discrete portions of IER, are feasible in an EDI system. According to Arar et al. (2014), each segment was largely made up of IER with the same polarity, such as AER or CER. Ionized species are eliminated in EDI in a manner similar to traditional ED, with the rate of ion elimination being significantly accelerated by the existence of the IER in the DLC (Dermentzis et al., 2012; Zheng et al., 2017). Improved transfer and electroregeneration are the two main purposes for EDI systems. The device's internal IER remains in salt forms in the improved transmission region. In this region, IE on IER surfaces serve as the transfer method. Electroregeneration is a term used to describe the second operational region for EDI devices. Continuous electrically driven water-splitting processes that regenerate IER are what make this region unique. To their hydrogen and hydroxide forms, the resins are "regenerated" (Meng et al., 2004; Zhang et al., 2022).

IERs are PS that may interact with ions that are free of a reverse charge because they retain their ionic locations. The exchange site for the ions in solution is provided by the ionic categories, or counter ions, of the IER. Counter ions can move through the polymeric material under a field of electricity because IERs in a solution of electrolyte are conductive to electricity, enabling the transfer of mass and the passage of related current. The movement and attraction of the opposing ions that the IERs are in touch with affect the electrical conductive properties of the IER. The electrical conductance in the IER bed is improved when the IER phase's opposing ion level exceeds than that of the solution's ion intensity. The IER phase in the DLC may consequently lower the electrical resistivity of the feed water, which is the primary benefit of employing IER in the EDI system (Yeon et al., 2003).

One of the key goals of many studies to lower energy usage is to minimize the electrical resistance of the component. The development of highly acidic and strongly basic IEM that can catalyze the process of water splitting and increase weak ion elimination, as well as the decrease of the CC's resistance, has been the main focuses of efforts to lower the total module impedance. Recirculating the concentrate fluid in a feed–and–bleed arrangement, injecting a saltwater solution into the concentrated stream, and adding IER to the CC are a few techniques for lowering CC impedance. In order to make the resistance free of the concentrated water conductance, it is better to reduce the module resistance by employing IER in the CC and EC in addition to the DLC (Wenten and Arfianto, 2013).

Combining IE and ED technologies revealed that the ion elimination was not an additive method. This combination technique filled IER in the DLC (Ho et al., 2010). The outcomes showed that the IER's only responsibility during the procedure was to keep the DLC conductive. The EDI with MIER was successful in removing the nitrate level of wastewater water from DLC that had increased quickly after electrification; it could be lowered to less than 10 ppm and met the WHO criteria. Therefore, it was demonstrated that using electricity to renew IER rather than chemicals was practical, and the IER's regeneration efficiency reached a maximum of as 60% after 90 hours.

4.2.2 Ion-exchange membrane

IEMs are among the most cutting-edge ones. The basic usages of the IEM process, which depends on the Donnan membrane, a state of equilibrium concept, have been focused on in order to address two significant ecological issues: the extraction and enrichment of useful ions, as well as the elimination of unsuitable ions from effluent water, particularly harmful metal ions from effluents (Goh et al., 2017; El Batouti et al., 2021). CEM only permits cations and blocks anions, whereas AEM only allows anions and prevents cations. According to Nagale et al. (2006), these kinds of membranes have the ability to be exploited as novel functional materials in the process of separation of ionic compounds, which typically occurs in solutions with multiple components.

IEM and IER have an identical chemical structure since they both contain functional groups (Hale et al., 1961). The physical demand of the membrane-making process is a major factor in the distinction between IEM and IER (Kononenko et al., 2017). However, AERs are soft, CERs are often brittle, and IERs are mechanically fragile. Due to their structure and method of manufacturing, the majority of commercially available IEM fall into one of two primary groups: homogeneous IEM or heterogeneous IEM (Luo et al., 2018).

Table 4.2 gives the IEM employed in different application of EDI.

IEM plays a crucial role in the EDI process since they can either accept or reject ions while creating DLC and CC. IEM serves to stop CC and DLC from mix in EDI. Excellent conductivity and permeability in IEM are sought. In contrast to a heterogeneous IEM, which uses two types of polymers to do those functions, a homogeneous IEM typically uses a single type of monomer to design the membrane's structure and

Table 4.2 Ion-exchange membrane employed in electrodeionization.

Application	Anion exchange membrane (AEM)	Cation exchange membrane (CEM)	References
Arsenic removal	Astom AMX AEM	Astom CMX CEM	Lee et al. (2017)
Boron removal	Neosepta AMX AEM	NeoseptaCMX CEM	Arar et al. (2013)
Boron removal	Selemion AHT AEM	Selemion CMD CEM	Arar et al. (2013)
Boron removal	Selemion AME AEM	Selemion CME CEM	Arar et al. (2013)
Calcium removal	Cationic 5B membrane	Anionic ACS membrane	Kurup et al. (2009)
Chromium removal	AMI-7001S Membranes International Inc.	CMI-7000 Membranes International Inc	Alvarado et al. (2013)
Chromium removal	Neosepta AFN	Neosepta CM-1	Alvarado et al. (2009)
Chromium removal	AEM, Double Flower Heterogeneous	CEM, Nafion117	Xing et al. (2009b)
Chromium removal	AMI-7001S Membranes International Inc.	CMI-7000 Membranes International Inc	Alvarado et al. (2015)
Chromium removal	AEM, Double Flower Heterogeneous	CEM, Nafion117	Xing et al. (2009a)
Chromium removal	—	CEM, Nafion117	Jina et al. (2020)
Cobalt removal	Neosepta AMX AEM	Neosepta CMX CEM	Yeon et al. (2004)
Copper removal	Neosepta AMX AEM	Neosepta CMX CEM	Arar et al. (2011)
Desalination	CMX strong acidic CEM	AMX strong basic AEM	Song et al. (2007)
Hardness removal	AMI-7001S	CMI-7000S	Zhang and Chen (2016)
Lithium removal	Selemion AHT	Selemion CMD	Demir et al. (2023)

Nitrate removal	AMI-7001S	CMI-7000S	Zhang and Chen (2016)
Nitrate removal	Neosepta AMX AEM	Neosepta CMX CEM	Yeon et al. (2004)
Production of pure water	Biopolar membrane	Biopolar membrane	Yeon et al. (2003)
Purification of organic acids	Neosepta AMX	Neosepta CMX	Lopez and Hestekin (2015)
Recovery of nickel	AMI-7001 AEM	CMI-7000 CEM	Dzyazko (2006)
Recovery of nickel	AMV AEM	CEM, Nafion117	Spoor et al. (2001)
Silica removal	Neosepta AMX AEM	Neosepta CMX CEM	Arar et al. (2013)
Silica removal	Selemion AHT AEM	Selemion CMD CEM	Arar et al. (2013)
Silica removal	Selemion AME AEM	Selemion CME CEM	Arar et al. (2013)
Sugar refining process	Anion, MA-3475	Cation, MC-3470	Widiasa et al. (2014)

supply charge groups. Because of this uneven distribution of functional groups, heterogeneous IEM exhibits reduced permselectivity and increased electrical resistance. Ion elimination, ion concentration, and ion extraction are therefore major applications of IEM procedures in WWT (Rathi and Kumar, 2020).

There are two routes in each stage of the unit, four DLC, and five CC overall. MIER is just filled with DLC. To facilitate nickel transfer in DLC, the volume ratio of CER:AER was 60:40. The total thickness of DLC and CC was 3 and 0.9 mm, accordingly, and the effective surface of each membrane was 162 cm^2. Additionally, the cell design has made certain improvements over the HPW production design. Between the cathode electrode and the final AEM next to it, a second CEM was introduced, resulting in the installation of a cell chamber known as the cathodic protection chamber. The cathodic protection chamber may be able to lessen nickel deoxidization on the cathode electrode and stop the cathode reaction's production of hydroxide ions from being transported to the CC. As a result, there was significantly less nickel hydroxide accumulation. To create an anodic protective chamber, an extra AEM was put between the AEM of the first DLC and the CEM next to the anode. The oxidative degradation to AEM of the first DLC, which results from anode reaction, may also be avoided, as well as the cationic transfer from anode chamber to the first DLC (Lu et al., 2010a,b,c).

4.2.3 Electrode

The CC and DLC are formed via the EDI system, which comprises of a sequence of AEM and CEM that are alternately positioned between two electrodes. Each compartment is filled with IER and the solution, which creates a voltage difference between the two electrodes. The cations are moved toward the cathode by the field of electricity, while the anions are propelled away from the cathode in the reverse direction and toward anode. Ions can only move according to the direction of their electrical charge because of the barriers created by the inserted IEM. One of the compartments gradually loses ions is DLC, whereas the other one gains ions is CC (Alvarado and Chen, 2014). Table 4.3 gives the electrodes employed in different application of EDI. The procedures taking place inside the electrode chambers, also known as EC, differ from those that happen inside the main chambers; therefore, an electrolyte solution is pumped through them. In conclusion, a weakly ionized substance may be

Table 4.3 Electrode employed in electrodeionization.

Application	Anode	Cathode	References
Chromium removal	$Ti/Ta_2O_5-IrO_2$	Stainless steel plate	Alvarado et al. (2013)
Chromium removal	$Ti/Ta_2O_5-IrO_2$	Stainless steel plate	Alvarado et al. (2009)
Chromium removal	$Ti/IrO_2-SnO_2-Sb_2O_5$	Titanium mesh	Xing et al. (2009a)
Sugar refining process	Platinum electrode	Stainless steel plate	Widiasa et al. (2014)
Nitrate removal	$Ti/Ta_2O_5-IrO_2$	Stainless steel plate	Zhang and Chen (2016)
Hardness removal	$Ti/Ta_2O_5-IrO_2$	Stainless steel plate	Zhang and Chen (2016)
Copper removal	Platinum Electrode	Stainless steel plate	Arar et al. (2011)
Nitrate removal	Ti/Ru	Ti/Ru	Bi et al. (2011)
Nickel removal	Ionic current sinks - graphite powder	Ionic current sinks - graphite powder	Dermentzis (2010)
Boron removal	Titanium with Ir/Ru coating	Stainless steel plate	Arar et al. (2013)
Silica removal	Titanium with Ir/Ru coating	Stainless steel plate	Arar et al. (2013)
Calcium removal	PTFE frame	Stainless steel plate	Kurup et al. (2009)
Arsenic removal	Platinum coated titanium	Stainless steel plate	Lee et al. (2017)
Copper removal	Platinized titanium grids	Platinized titanium grids	Dermentzis et al. (2009)
Chromium removal	$Ti/Ta_2O_5-IrO_2$	Stainless steel plate	Alvarado et al. (2015)
Chromium removal	$Ti/IrO_2-SnO_2-Sb_2O_5$	Titanium mesh	Xing et al. (2009a)
Recovery of nickel	Platinum Electrode	Platinum Electrode	Dzyazko (2006)
Hardness removal	Ruthenium-coated titanium	Stainless steel plate	Fu et al. (2009)
Lithium removal	Noble	Stainless steel plate	Demir et al. (2023)
High pure water production	Stable electrode	Stainless steel plate	Jordan et al. (2020)
Cesium removal	Stainless steel plate	Stainless steel plate	Zahakifar et al. (2017)
Strontium removal	Stainless steel plate	Stainless steel plate	Zahakifar et al. (2017)
Chromium removal	$Ti/IrO_2-SnO_2-Sb_2O_5$	Titanium mesh	Jina et al. (2020)

reliably and consistently eliminated through an EDI system. The resins' lifespan is significantly increased and pH shock is also completely eliminated by the continuous renewal procedure (Kumar et al., 2022/2022).

4.3 Types of electrodeionization

EDI comprises of IEM, IER, and electrodes. EDI can be classified into three types based on the module design. They are (1) plate and frame EDI, (2) spiral wound EDI, and (3) stacked disk stacks EDI. It also classified based on the recent advances. They are resin wafer EDI, electrostatic shielding EDI, EDI reversal, and membrane-free EDI. Based on the arrangement of IER, they are classified into three types including anion exchange EDI, cation exchange EDI, and mixed bed EDI.

4.3.1 Anion exchange electrodeionization

Anion exchange EDI stacks whereby the DLC is made up of two AEM and the AER have been filled in the DLC (Fig. 4.1A). In some case, anion exchange EDI stacks whereby the DLC is made up of AEM and

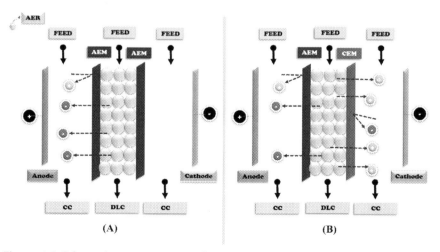

Figure 4.1 Schematic representation of anion exchange electrodeionization (A) two AEM filled with AER (B) One AEM and one CEM filled in between with AER. *AEM*, Anion exchange membrane; *AER*, anion-exchange resin; *CEM*, cation exchange membrane.

CEM and the AER have been filled in the DLC (Fig. 4.1B). Anode, cathode, and center chambers were the three distinct chambers that made up the EDI unit. The center chamber of the AER was full. An AEM and CEM were used to divide the EC from the center chamber. The effluent from the synthetic process, which was physically recycled, is where the anolyte came from. The catholyte was a 0.5 M H_2SO_4 solution that was not regenerated in order to boost conductance. It was determined that the gel strong-base IER would work best for EDI. It was demonstrated that EDI could efficiently extract and reclaim Cr(VI) from effluent. A pure chromic acid mixture comprising Cr(VI) at a level of up to 6300 mg/L Cr(VI) was produced after the treatment, which decreased the Cr(VI) level from the initial 40−0.09 ppm. According to Xing et al. (2009b), the current efficacy and consumption of energy were 16.1% and 4.1 kWh/mol chromium, correspondingly.

Anode, cathode, and a DLC for chromium elimination made up the three individual compartments that made up the EDI device. As a CC, the anode chamber also performed. The IER served as the ion exchanger and was situated in the DLC. An AEM and a CEM were used for dividing the EC and the DLC. The effective area of both membranes was 40 cm^2. Between AEM and CEM, there was a 9-mm net space. The catholyte utilized for every EDI study was 0.5 L of a 0.5 M sulfuric acid solution, and the anolyte used in each experiment was 0.5 L of manually regenerated artificial wastewater (Xing et al., 2009a). The conduction of ions was only possible for anions via the solid phase when AERs were used in the DLC. Cations and the remainder of anions were transported via the aqueous phase, leading to an imbalance in the ionic transportation velocities.

4.3.2 Cation exchange electrodeionization

Cation exchange EDI stacks whereby the DLC is made up of two CEM and the CER have been filled in the DLC (Fig. 4.2A). In some case, anion exchange EDI stacks whereby the DLC is made up of AEM and CEM and the CER have been filled in the DLC (Fig. 4.2B). A three-compartment cell with a dimensionally stable anode, cathode, and three different stream lines makes up the test apparatus for removing copper. With the help of the AEM and CEM, the cell's CC were isolated from the center compartment. The potent acid CER was poured into the center container. The EDI cell's primary measurements are 0.12 m in height,

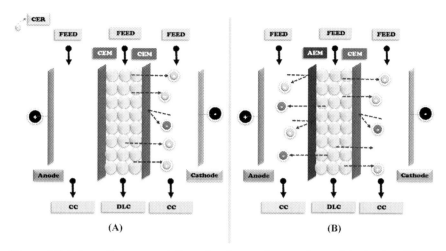

Figure 4.2 Schematic representation of cation exchange electrodeionization (A) two CEM filled with CER (B) One AEM and one CEM filled in between with CER. *AEM*, Anion exchange membrane; *CEM*, cation exchange membrane; *CER*, cation-exchange resin.

0.07 m in width, and 0.038 m in length. The cell's internal membranes have an effective area of 10.2 cm^2. A magnetic pump was used to move a 1 L of copper sulfate solution via the center chamber. Sulfuric acid solution diluted to 1 L was passed via the EC. The samples (10 cm^3) were collected at periodic times from the EC, DLC, and CC streams and then subjected to an AAS analysis to determine the level of copper ions (Arar et al., 2011).

The CEM, AEM, or sheets made of inert polyethylene were used to divide the cell's center chamber from the EC. The IEM was put between a bed of a nickel-loaded CER. Platinum electrodes had an effective area of 15 cm^2 and membranes had the same size. The center chamber underwent a "once-through" procedure with either deionized water as the medium or the nickel sulfate solution. Utilizing liquid pressure gauges at the intake and outflow of the center chamber, a pressure decrease across the CER bed was managed. The EC were either left unfilled or filled with sulfuric acid solutions. According to Dzyazko (2006), both the catholyte and anolyte had similar volumes. The conduction of ions was only possible for cations via the solid phase when CERs were used in the DLC. Anions and the remainder of cations were transported via the aqueous phase, leading to an imbalance in the ionic transportation velocities.

4.3.3 Mixed bed electrodeionization

Mixed bed EDI stacks whereby the DLC is made up of AEM and CEM and the MIER have been filled in the DLC (Fig. 4.3). The procedure is now much more efficient because of a number of improvements to the initial EDI stack construction that have been made in the past few years. Loading the CC with an IER considerably decreased the electrical conductivity of the stack. By substituting clustered or stacked beds of CER and AER for the MIER in the DLC, several drawbacks of the MIER were offset, including the ineffective removal of weak acids (Lu et al., 2010a,b,c). Even if the elimination of weak acids was greatly enhanced in EDI stacks with clustered or stacked beds, a different issue still persists. The current passing through the CER and AER, and consequently, their regenerative rate, varies because the CER and AER have differing conductance. In an EDI stack with entirely distinct CER and AER beds, the issue of varying restoration speeds or flow rates in the CER and the AER can be solved by making use of the electrodes' ability to produce the hydrogen and hydroxyl ions required for the renewal of IER (Alvarado et al., 2013). Combining IE and ED techniques revealed that the ion elimination is not a sequential process. The outcomes show

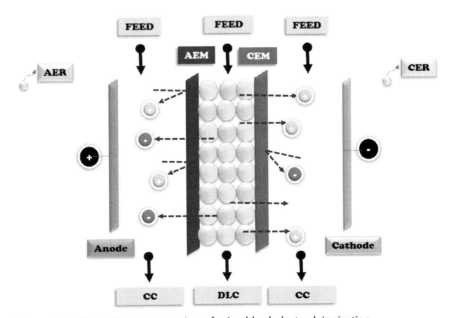

Figure 4.3 Schematic representation of mixed bed electrodeionization.

that the IER's only responsibility is to keep the DLC conductive throughout the procedure. But when a MIER is employed, the EDI permits a maximum elimination in a shorter amount of time primarily because the number of interaction sites between AER and CER has increased (Alvarado et al., 2009).

EDI system is used in the process of refining sugar to remove inorganic impurities. With MIER, the DLC of the EDI stack was filled. After then, investigations were run in batch and continuous modes. The outcomes shown that EDI can eliminate color body and inorganic impurities from sugar solutions. Major ion elimination is seen to significantly diminish at reasonably high levels of sugar. Current efficacy and pH were reduced as a result of the increased current density and voltage that was used, which had minimal effect on the elimination of ions. According to Widiasa et al. (2014), the color disappearance may be related to complicated bond breakdown caused by dissociation of water within the DLC.

The EDI stack configuration was designed as a series of transparent sheets of acrylic and IEM among the anode and cathode to divide the CC and DLC for the purpose of eliminating nitrate and hardness. IEM with an effective area of 60 cm^2 and two sheets of acrylic housing the two electrodes to construct the EC were encased among the sheets of acrylic. In total, 20 g of MIER was placed in the DLC in the space between the CEM and AEMs. To evenly disperse the pressure, rubber gasket separators were placed on either side of the IEM. Additional equipment for the EDI stack included pumps, tanks, power source, and a flow meter. The DLC, CC, and EC were found to have thicknesses of around 0.01 m, 0.01 mm, and 0.015 mm, correspondingly, in this EDI setup. The DLC was surrounded by the EC and the cation CC and anion CC were placed on both sides of the DLC. In all of the chambers that made up the EDI cell, a constant flow of solution moved from the bottom to the top. Prior to the implementation of the EDI, CER had become saturated with magnesium and calcium ions, whereas AERs were saturated with nitrate ions. This was done because both unsaturated and saturated IER have differing effects on how well the EDI functions. Once the IE equilibrium of the IER was established, the EDI tests were conducted several times unless no further decline in the cell voltage could be seen while the exact same applied current was being used (Zhang and Chen, 2016).

For copper elimination, it was a two-stage stack with three cell pairings in each step. A CEM and an AEM made up the DLC. The effective area of each membrane was 135 cm^2. The DLC had a 3 mm thickness. In the DLC, a blend of CER and AER was employed. The CC had a 0.8 mm thickness. Nonconducting, finely woven screens were used for filling the CC (Guan and Wang, 2007). Arsenic removal required the placement of CC on both ends of the DLC, which also served as the EC. The DLC had a 7.2 mL capacity. The DLC was jam-packed with IER. A series of IEM between the electrodes served as the partitions between each chamber. The effective membrane area of each EDI cell was 24 cm^2. Gaskets dispersed the process liquids among the chambers in addition to supporting the membranes that surrounded them. The EDI analysis was performed using a MIER. The entire exchange capability was a 1:1 mixture of CER and AER. In order to determine the features of a brand-new IER that has not undergone the procedure of regeneration, the fresh unit was studied. The IER is a potent adsorbent by itself. Even in the absence of an electric field, unsaturated fresh IER might adsorb the inflowing ions (Lee et al., 2017).

With the CER bed in front of both the AER bed and MIER bed, the multilayered bed-packed EDI device was performed. Three layers made up the IER bed: the bottom layer featured a CER for HM elimination, the center layer an AER for anion elimination, and the upper layer a MIER for pH regulation. Over 99% of the ions were eliminated by this stack structure, which also had a 30% current efficiency, by preventing an interaction between the HM and hydroxide ions. The three-stage EDI system that was mimicked by the layered-bed EDI device had an under-control wastewater pH. A HPW may be attained in a real process using multistage EDI (Yeon et al., 2004).

4.4 Chemistry in electrodeionization

The CEM and AEM are positioned alternately between the anode and the cathode in order to provide DLC and CC in an EDI cell setup. In an EDI stack, the highly concentrated solution is pumped via the IER, that is sandwiched among the CEM and AEM (Pan et al., 2017). This allows ions that are positive to flow through the CEM while repelling negative ions from the AEM, and vice versa. Alternating positions are used for the AEM and CEM. From that point on, the diluted solution has been currently gathered in the different DLC (Otero et al., 2021). IER beds are introduced into the DLC to provide a substrate for electrical activity that prevents the growth of the concentration polarization effect. In order to facilitate mass movement through IER, the applied electric field promotes ionic movement via IER bed in two distinct chambers that absorb and eject the ionic particles (Saravanan et al., 2023; Song et al., 2005). The general chemical reaction happening in electrode and IER with A^+ (cation) and B^- (anion) is given below,

Chemical reaction at electrode
Anode:

$$2H_2O \Leftrightarrow 4H^+ + O_2 + 4e^- \tag{4.1}$$

Cathode:

$$2H_2O + 2e^- \Leftrightarrow H_2 + 2OH^- \tag{4.2}$$

Chemical reaction at IER
AER:

$$R - H + A^+B^- \rightarrow R - A + HB \tag{4.3}$$

CER:

$$R - OH + A^+B^- \rightarrow R - B + AOH^- \tag{4.4}$$

MIER:

$$R - H + A^+B^- \rightarrow R - A + HB \tag{4.5}$$

$$R - OH + A^+B^- \rightarrow R - B + AOH^- \tag{4.6}$$

IER regeneration:

$$H_2O \rightarrow H^+ + OH^- \tag{4.7}$$

$$R - A + H^+ \rightarrow R - H + A^+ \tag{4.8}$$

$$R - B + OH^- \rightarrow R - OH + B^- \tag{4.9}$$

4.5 Conclusion

In this chapter, the primary elements of the EDI unit, such as IER, IEMs, and electrodes, were explored. Also, the various types of EDI including anion exchange EDI, cation exchange EDI, and mixed bed EDI and the chemistry involved in the EDI procedure (anode, cathode, AER, CER, MIER, and IER regeneration) were discussed. From solutions containing metal ions, HPW was created using the EDI procedure. Without the necessity of chemical regenerants, the multilayered bed EDI design can eliminate HM ions at high purity levels from industrial plant wastes including plating wastewater and nuclear primary coolants. A mixed bed EDI is used in place of anion exchange or cation exchange, giving anions and cations access to the solid pathway. As a result, the resulting conductivity will be significantly better than the conductance that would be produced in solution, and the charge transfer speed will be regulated. On the other side, mixed beds provide more locations for in situ renewal procedures, which helps systems using EDI run more quickly.

References

Alvarado, L., Chen, A., 2014. Electrodeionization: principles, strategies and applications. Electrochimica Acta 132, 583–597.

Alvarado, L., Ramírez, A., Rodríguez-Torres, I., 2009. Cr(VI) removal by continuous electrodeionization: study of its basic technologies. Desalination 249 (1), 423–428.

Alvarado, L., Torres, I.R., Chen, A., 2013. Integration of ion exchange and electrodeionization as a new approach for the continuous treatment of hexavalent chromium wastewater. Separation and Purification Technology 105, 55–62.

Alvarado, L., Rodríguez-Torres, I., Balderas, P., 2015. Investigation of current routes in electrodeionization system resin beds during chromium removal. Electrochimica Acta 182, 763–768.

Arar, Ö., Yüksel, Ü., Kabay, N., Yüksel, M., 2011. Removal of Cu2 + ions by a microflow electrodeionization (EDI) system. Desalination (1–3), 296–300. Available from: https://doi.org/10.1016/j.desal.2011.04.044.

Arar, O., Yüksel, U., Kabay, N., Yüksel, M., 2013. Application of electrodeionization (EDI) for removal of boron and silica from reverse osmosis (RO) permeate of geothermal water. Desalination. 25–33. Available from: https://doi.org/10.1016/j.desal.2012.10.001.

Arar, O., Yüksel, U., Kabay, N., Yüksel, M., 2014. Various applications of electrodeionization (EDI) method for water treatment—a short review. Desalination. 16–22. Available from: https://doi.org/10.1016/j.desal.2014.01.028.

Bajpai, P., 2018. Biermann's Handbook of Pulp and Paper.

Bi, J., Peng, C., Xu, H., Ahmed, A.S., 2011. Removal of nitrate from groundwater using the technology of electrodialysis and electrodeionization. Desalination and Water Treatment (1–3), 394–401. Available from: https://doi.org/10.5004/dwt.2011.2891.

Bresser, D., Buchholz, D., Moretti, A., Varzi, A., Passerini, S., 2018. Alternative binders for sustainable electrochemical energy storage — the transition to aqueous electrode processing and bio-derived polymers. Energy & Environmental Science 11, 3096−3127. Available from: https://doi.org/10.1039/c8ee00640g.

Demir, G., Mert, A.N., Arar, O., 2023. Utilization of Electrodeionization for Lithium Removal. ACS Omega.

Dermentzis, K., 2010. Removal of nickel from electroplating rinse waters using electrostatic shielding electrodialysis/electrodeionization. Journal of Hazardous Materials 173 (1−3), 647−652.

Dermentzis, K., Davidis, A., Papadopoulou, D., Christoforidis, A., Ouzounis, K., 2009. Copper removal from industrial wastewaters by means of electrostatic shielding driven electrodeionization. Journal of Engineering Science and Technology Review 1, 131−136. Available from: https://doi.org/10.25103/jestr.021.24.

Dermentzis, K., Davidis, A., Chatzichristou, C., Dermentzi, A., 2012. Ammonia removal from fertilizer plant effluents by a coupled electrostatic shielding based electrodialysis/ electrodeionization process. Global Nest Journal 4, 468−476. Available from: https:// doi.org/10.30955/gnj.000703, http://www.gnest.org/journal/Vol_14_no_4/468-476_703_Dermentzis_14-4.pdf.

Dzyazko, Y.S., 2006. Purification of a diluted solution containing nickel using electrodeionization. Desalination (1−3), 47−55. Available from: https://doi.org/10.1016/j. desal.2006.09.008.

El Batouti, M., Al-Harby, N.F., Elewa, M.M., 2021. A review on promising membrane technology approaches for heavy metal removal from water and wastewater to solve water crisis. Water 13 (22), 3241.

Fu, L., Wang, J., Su, Y., 2009. Removal of low concentrations of hardness ions from aqueous solutions using electrodeionization process. Separation and Purification Technology 68 (3), 390−396.

Gifford, J.D., Atnoor, D.E.V.E.N., 2000, October. An innovative approach to continuous electrodeionization module and system design for power applications. In: International Water Conference. Pittsburg, P A. Oetober, pp. 22−26.

Goh, P.S., Matsuura, T., Ismail, A.F., Ng, B.C., 2017. The Water−Energy Nexus: Solutions towards Energy-Efficient Desalination. Energy Technology 8, 1136−1155. Available from: https://doi.org/10.1002/ente.201600703, http://onlinelibrary.wiley. com/journal/10.1002/(ISSN)2194-4296.

Guan, S., Wang, S., 2007. Experimental studies on electrodeionization for the removal of copper ions from dilute solutions. Separation Science and Technology 42 (5), 949−961.

Hakim, A.N., Khoiruddin, K., Ariono, D., Wenten, I.G., 2020. Ionic separation in electrodeionization system: mass transfer mechanism and factor affecting separation performance. Separation & Purification Reviews 4, 294−316. Available from: https://doi. org/10.1080/15422119.2019.1608562.

Hakim, A.N., Wenten, I.G., 2014. Advances in electrodeionization technology for ionic separation—a review. Membrane Water Treatment 5 (2), 087.

Hale, D.K., McCauley, D.J., 1961. Structure and properties of heterogeneous cation-exchange membranes. Transactions of the Faraday Society 57, 135−149.

Ho, T., Kurup, A., Davis, T., Hestekin, J., 2010. Wafer chemistry and properties for ion removal by wafer enhanced electrodeionization. Separation Science and Technology 4, 433−446. Available from: https://doi.org/10.1080/01496390903526709.

Hu, J., Chen, Y., Zhu, L., Qian, Z., Chen, X., 2016. Production of high purity water using membrane-free electrodeionization with improved resin layer structure. Separation and Purification Technology 89−96. Available from: https://doi.org/ 10.1016/j.seppur.2016.03.027, http://www.journals.elsevier.com/separation-and-purification-technology/.

Hu, J., Fang, Z., Jiang, X., Li, T., Chen, X., 2015. Membrane-free electrodeionization using strong-type resins for high purity water production. Separation and Purification Technology 90−96. Available from: https://doi.org/10.1016/j.seppur.2015.02.023, http://www.journals.elsevier.com/separation-and-purification-technology/.

Jina, Q., Yaob, W., Chenc, X., 2020. Removal of Cr (VI) from wastewater by simplified electrodeionization. Desalination and Water Treatment 183, 301−306.

Jordan, M.L., Valentino, L., Nazyrynbekova, N., Palakkal, V.M., Kole, S., Bhattacharya, D., et al., 2020. Promoting water-splitting in Janus bipolar ion-exchange resin wafers for electrodeionization. Molecular Systems Design and Engineering 5, 922−935. Available from: https://doi.org/10.1039/c9me00179d, http://www.rsc.org/journals-books-databases/about-journals/molecular-systems-design-engineering/.

Kononenko, N., Nikonenko, V., Grande, D., Larchet, C., Dammak, L., Fomenko, M., et al., 2017. Porous structure of ion exchange membranes investigated by various techniques. Advances in Colloid and Interface Science 196−216. Available from: https://doi.org/10.1016/j.cis.2017.05.007.

Kumar, P.S., Varsha, M., Rathi, B.S., Rangasamy, G., 2022. Electrodeionization: fundamentals, methods and applications. Environmental Research.

Kurup, A.S., Ho, T., Hestekin, J.A., 2009. Simulation and optimal design of electrodeionization process: separation of multicomponent electrolyte solution. Industrial & Engineering Chemistry Research 48 (20), 9268−9277.

Lee, D., Lee, J.Y., Kim, Y., Moon, S.H., 2017. Investigation of the performance determinants in the treatment of arsenic-contaminated water by continuous electrodeionization. Separation and Purification Technology 381−392. Available from: https://doi.org/10.1016/j.seppur.2017.02.025, http://www.journals.elsevier.com/separation-and-purification-technology/.

Li, F.U.Z., Zhang, M., Zhao, X., Hou, T., Liu, L.I.J., 2010. Removal of Co^{2+} and Sr^{2+} from a primary coolant by continuous electrodeionization packed with weak base anion exchange resin. Nuclear Technology 1, 71−76. Available from: https://doi.org/10.13182/NT10-A10883, http://www.new.ans.org/pubs/journals/nt/.

Lopez, A.M., Hestekin, J.A., 2015. Improved organic acid purification through wafer enhanced electrodeionization utilizing ionic liquids. Journal of Membrane Science 493, 200−205.

Lu, H., Wang, J., Yan, B., Bu, S., 2010a. Recovery of nickel ions from simulated electroplating rinse water by electrodeionization process. Water Science and Technology 61 (3), 729−735.

Lu, J., Wang, Y.X., Zhu, J., 2010b. Numerical simulation of the electrodeionization (EDI) process accounting for water dissociation. Electrochimica Acta 55 (8), 2673−2686.

Lu, J., Wang, Y.X., Lu, Y.Y., Wang, G.L., Kong, L., Zhu, J., 2010c. Numerical simulation of the electrodeionization (EDI) process for producing ultrapure water. Electrochimica Acta 55 (24), 7188−7198.

Lu, H., Wang, Y., Wang, J., 2015. Recovery of Ni^{2+} and pure water from electroplating rinse wastewater by an integrated two-stage electrodeionization process. Journal of Cleaner Production 92, 257−266.

Luo, T., Abdu, S., Wessling, M., 2018. Selectivity of ion exchange membranes: a review. Journal of Membrane Science 555, 429−454.

Meng, H., Peng, C., Song, S., Deng, D., 2004. Electro-regeneration mechanism of ion-exchange resins in electrodeionization. Surface Review and Letters 6, 599−605. Available from: https://doi.org/10.1142/S0218625X04006542.

Nagarale, R.K., Gohil, G.S., Shahi, V.K., 2006. Recent developments on ion-exchange membranes and electro-membrane processes. Advances in Colloid and Interface Science 119 (2−3), 97−130.

Nikonenko, V., Nebavsky, A., Mareev, S., Kovalenko, A., Urtenov, M., Pourcelly, G., 2018. Modelling of ion transport in electromembrane systems: Impacts of membrane bulk and surface heterogeneity. Applied Sciences 9 (1), 25.

Otero, C., Urbina, A., Rivero, E.P., Rodríguez, F.A., 2021. Desalination of brackish water by electrodeionization: experimental study and mathematical modeling. Desalination. Available from: https://doi.org/10.1016/j.desal.2020.114803.

Pan, S.Y., Snyder, S.W., Ma, H.W., Lin, Y.J., Chiang, P.C., 2017. Development of a resin wafer electrodeionization process for impaired water desalination with high energy efficiency and productivity. ACS Sustainable Chemistry and Engineering 4, 2942−2948. Available from: https://doi.org/10.1021/acssuschemeng.6b02455, http://pubs.acs.org/journal/ascecg.

Rathi, B.S., Kumar, P.S., 2020. Electrodeionization theory, mechanism and environmental applications. A review. Environmental Chemistry Letters 18, 1209−1227.

Rathi, B.S., Kumar, P.S., Parthiban, R., 2022. A review on recent advances in electro-deionization for various environmental applications. Chemosphere 289, 133223.

Saravanan, A., Yaashikaa, P.R., Senthil Kumar, P., Karishma, S., Thamarai, P., Deivayanai, V.C., Rangasamy, G., Selvasembian, R., Aminabhavi, T.M., 2023. Environmental sustainability of toxic arsenic ions removal from wastewater using elec-trodeionization. Separation and Purification Technology. Available from: https://doi.org/10.1016/j.seppur.2023.123897.

Sarıçiçek, E.N., Tuğaç, M.M., Özdemir, V.T., İpek, İ.Y., Arar, Ö., 2021. Removal of boron by boron selective resin-filled electrodeionization. Environmental Technology and Innovation. Available from: https://doi.org/10.1016/j.eti.2021.101742, http://www.journals.elsevier.com/environmental-technology-and-innovation/.

Shen, X., Yu, J., Chen, Y., Peng, Z., Li, H., Kuang, X., et al., 2019. Influence of some parameters on membrane-free electrodeionization for the purification of wastewater containing nickel ions. International Journal of Electrochemical Science 11237−11252. Available from: https://doi.org/10.20964/2019.12.35, http://www.electrochemsci.org/abstracts/vol14/141211237.pdf.

Skogley, E.O., Dobermann, A., 1996. Synthetic ion-exchange resins: soil and environ-mental studies. Journal of Environmental Quality 25 (1), 13−24.

Song, J.H., Yeon, K.H., Moon, S.H., 2005. Transport characteristics of Co^{2+} through an ion exchange textile in a continuous electrodeionization (CEDI) system under elec-tro-regeneration. Separation Science and Technology 39 (15), 3601−3619.

Song, J.H., Yeon, K.H., Moon, S.H., 2007. Effect of current density on ionic transport and water dissociation phenomena in a continuous electrodeionization (CEDI). Journal of Membrane Science 291 (1−2), 165−171.

Spoor, P.B., Veen, W.T., Janssen, L.J.J., 2001. Electrodeionization 1: migration of nickel ions absorbed in a rigid, macroporous cation-exchange resin. Journal of Applied Electrochemistry 31, 523−530.

Vagliasindi, F.G., Belgiorno, V., Napoli, R.M., 1998. Water treatment in remote and rural areas: a conceptual screening protocol for appropriate pou/poe technologies. Environmental Engineering and Renewable Energy. Elsevier, pp. 329−336.

Wenten, I.G., Arfianto, F., 2013. Bench scale electrodeionization for high pressure boiler feed water. Desalination 314, 109−114.

Widiasa, I.N., Wenten, I.G., 2014. Removal of inorganic contaminants in sugar refining process using electrodeionization. Journal of Food Engineering 133, 40−45.

Wood, J., 2008. Continuous electrodeionization for water treatment at power plants. Power Engineering 112 (9), 62−65.

Wood, J., Gifford, J., 2003. Boiler feedwater treatment: improvements in continuous elec-trodeionization. Power Engineering (Barrington, Illinois) 9, 42−46.

Wood, J., Gifford, J., Arba, J., Shaw, M., 2010. Production of ultrapure water by continuous electrodeionization. Desalination 3, 973−976. Available from: https://doi.org/10.1016/j.desal.2009.09.084.

Xing, Y., Chen, X., Wang, D., 2009a. Variable effects on the performance of continuous electrodeionization for the removal of Cr (VI) from wastewater. Separation and Purification Technology 68 (3), 357−362.

Xing, Y., Chen, X., Yao, P., Wang, D., 2009b. Continuous electrodeionization for removal and recovery of Cr(VI) from wastewater. Separation and Purification Technology 2, 123−126. Available from: https://doi.org/10.1016/j.seppur.2009.03.029.

Xu, T., 2005. Ion exchange membranes: state of their development and perspective. Journal of Membrane Science 263 (1-2), 1−29.

Yeon, K.H., Moon, S.H., 2003. A study on removal of cobalt from a primary coolant by continuous electrodeionization with various conducting spacers. Separation Science and Technology 38 (10), 2347−2371.

Yeon, K.H., Seong, J.H., Rengaraj, S., Moon, S.H., 2003. Electrochemical characterization of ion-exchange resin beds and removal of cobalt by electrodeionization for high purity water production. Separation Science and Technology 2, 443−462. Available from: https://doi.org/10.1081/SS-120016584.

Yeon, K.H., Song, J.H., Moon, S.H., 2004. A study on stack configuration of continuous electrodeionization for removal of heavy metal ions from the primary coolant of a nuclear power plant. Water Research 38 (7), 1911−1921.

Zahakifar, F., Keshtkar, A., Nazemi, E., Zaheri, A., 2017. Optimization of operational conditions in continuous electrodeionization method for maximizing Strontium and Cesium removal from aqueous solutions using artificial neural network. Radiochimica Acta 7, 583−591. Available from: https://doi.org/10.1515/ract-2016-2709, http://www.degruyter.com/view/j/ract.2013.101.issue-7/ract.2013.101.issue-7/ract.2013.101.issue-7.xml.

Zahakifar, F., Keshtkar, A.R., Souderjani, E.Z., Moosavian, M.A., 2020. Use of response surface methodology for optimization of thorium(IV) removal from aqueous solutions by electrodeionization (EDI). Progress in Nuclear Energy. Available from: https://doi.org/10.1016/j.pnucene.2020.103335.

Zhang, Z., Chen, A., 2016. Simultaneous removal of nitrate and hardness ions from groundwater using electrodeionization. Separation and Purification Technology 164, 107−113.

Zheng, H., Gong, X., Yang, Y., Yang, J., Yang, X., Wu, Z., 2017. Concentration of nitrogen as new energy source from wastewater by electrodeionization. Energy Procedia, Elsevier Ltd, China, pp. 1421−1426, 142. http://www.sciencedirect.com/science/journal/18766102.

Zhang, X., Jin, H., Deng, S., Xie, F., Li, S., Chen, X., 2022. Amphoteric blend ion exchange resin with medium-strength alkalinity for high-purity water production in membrane-free electrodeionization. Desalination. Available from: https://doi.org/10.1016/j.desal.2022.115663.

CHAPTER FIVE

Application and comparison of electrodeionization

5.1 Introduction

Ion exchange (IE) and membrane technology (MT) have traditionally been used in conjunction to produce high purity water (HPW). In order to create an extremely efficient demineralization procedure, electrodeionization (EDI) integrates semiimpermeable MT with IE medium (Alvarado and Chen 2014). The fundamental design of an EDI stack is similar to that of a deionization processes chamber. A cationic exchange membrane (CEM) and an anionic exchange membrane (AEM) are sandwiched in a vessel within an ion-exchange resin (IER). Water cannot enter across the membrane; ions alone may. A direct current (DC) is used to drive the IER in the central compartment (CC), which improves the transfer of cations or anions. Hydrogen and hydroxyl ions are produced during EDI operation via the water disintegration process. Without the need of regeneration chemicals, generated hydrogen and hydroxyl ions constantly renew the IER electrically (Arar et al., 2014).

In contrast to IE beds that are used in batch processing, the IE beds in EDI systems are continually renewed to prevent exhaustion. The concentrates used to eliminate the impurities prevent them from accumulating and wearing down the IER. It may be several years before an EDI unit has to be replaced. Typically, this method yields products with water resistivities of >15 MΩ cm. Single-use cleaning cartridges can be replaced with this technique (Rathi & Kumar, 2020). Any application that calls for the efficient and continuous elimination of water pollutants without the need for hazardous chemicals can benefit from EDI. Reusing leftover water in the food and beverages business, chemical manufacturing, the field of biotechnology gadgets, cosmetics, laboratories, pharmaceutical industry (PI), boiler water for feed, and lowering the amount of organic matter are a few applications (Rathi et al., 2022).

Conventional techniques include adsorption, MT, IE, and electrochemical technologies including ED and EDI (Shrestha et al., 2021). When the heavy metals (HMs) are dissolved in significant amounts of a solution at a relatively low level, the majority of these approaches may be prohibitively costly or useless. EDI is one of the most extensively employed ways among all those put out, and it is also regarded as an efficient, practical, and cost-effective technology for wastewater treatment (WWT) (Razzak et al., 2022). No pollutant accumulation, chemical-free, ion-free final result, and recycling are a few of the key advantages of EDI chemical removal. This chapter's goal was to investigate EDI and its applications in many fields, such as hydrometallurgy, electroplating, mining, and pharmaceuticals. Adsorption, reverse osmosis (RO), the IE process, membrane filtration, and ED are all examined in detail, and their costs and energy requirements are contrasted with EDI.

5.2 Electrodeionization

The hybrid IE/ED treatment, often known as EDI, has been demonstrated to be a very effective method for desalinating seawater with significant environmental benefits. EDI enables the two procedures' respective shortcomings, ED and IE. As water is desalinated, ED exhibits increased electric resistance in the dilute compartment (DLC). As a result, additional power is needed to create HPW, which significantly reduces efficiency. On the other hand, because the IE process has to chemically regenerate spent IER in order to reuse it, it cannot operate continually. However, the EDI method enables continuous operation under the impact of the electric field to renew the IER owing to the combination of two techniques. As a result, steady water quality may be generated. Without the requirement for chemical IER renewal, the electrical current placed between the two electrodes constantly desalinates water. Due to this significant benefit, EDI is an environmentally beneficial procedure (Otero et al., 2021).

Permselective AEM and CEM are often placed in alternate series among a pair of electrodes in an EDI unit. A separator is put among a pair of AEM and CEM in order to construct a chamber. An EDI stack has three different types of chambers: electrode chambers (ECs), DLC, and CC. With IER, each of the compartments are packed. By raising the total stack electrical conductivity, the IER contributes to improving the transit of ionized substances from the bulk solution toward the IEM. A plate-and-frame unit contains the majority of commercial EDI installations. Spiral wound and stacked-disk stacks are two other stack forms that are accessible (Hakim et al., 2020). The key advantage of EDI is better water quality that is accessible for less money to operate. EDI also has the advantage of producing less effluent. Higher rates of flow, where gallons per minute may be collected and returned to the unprocessed water from the breakdown tank, can make this particularly significant. The initial treatment efficiency is actually improved in this situation since the EDI concentrated water is of higher quality than the unprocessed water supply, allowing for successful operation at a 100% recovery rate (Peng and Guo 2020).

When compared to traditional systems of ionic exchange, the EDI has a number of benefits, including low power usage, affordable maintenance and operation, a minimal need for complicated regulate patterns requiring operator oversight, a small footprint demand, and complete elimination of dissolved inorganic particulates. For water with a hardness greater than 1, EDI is not suitable because calcium carbonate that is present would form a scab in the camera of the concentrated water, restricting the process. It additionally needs prior to treatment for cleansing and to keep an efficient system running, the membranes that allow the transferred ions travel will occasionally need substitution, which tends to be costly (Tate 2014).

5.3 Application of electrodeionization

Deionized water, or DI water, is produced by the EDI process, and it is beneficial in a variety of industries and applications. Electronic devices, testing in laboratory amenities, production, PI, electrical power generation and cooling down food processing, and manufacturing of chemicals are a few of the sectors that require clean, deionized water for their operations. The practical examples provided have demonstrated the effectiveness of the EDI method in protecting the surroundings, producing HPW, and recovering certain valuable species. Normal MT such as RO and electrodialysis reversal (EDIR) has trouble removing weakly-ionized ions such as carbon dioxide and boron. In a highly significant degree, the EDI benefits from constant elimination of these kinds of species (Kumar et al., 2022).

5.3.1 Mining industry

The majority of the resources we need to construct our everyday tools and facilities to create enormous quantities of energy and to generate the majority of our food are provided by the mining industries (MIs). At the same, a period of time MI is the activity of humans that has had the greatest negative environmental effects and is associated with significant social repercussions and inequities. Nevertheless, MI is crucial to the future of our nation (Carvalho 2017). The general public's concern for the food preservation, security, and rehabilitation of the natural environment has grown in the past few years. Around the globe, legislation has been done on the issues related to the environment that have an impact on the MI in this manner. In this fresh structure, two issues—project evaluation using social cost—benefit analysis and environmental damage assessments—are of utmost relevance to the industry. Both challenges often call for study to be done in a fashion that allows the depletion of resources from the environment and the impacts caused by environmental deterioration on the well-being and health of people to be quantified (Damigos 2006).

The need for minerals, metals, and micronutrients has significantly expanded as a result of economic expansion and recent technological breakthroughs. As a result, there is a lot of interest in creating technology to extract these materials from seawater, other brines, and streams of wastewater. In addition to enhancing their continuous supply, the recuperation of metals from these aqueous sources—commonly referred to as "water mining"—also renders the main process cleaner and more cost-effective by lowering the ecological footprint and turning "wastage" into product. In recent years, the environmental issues in the MI have received a lot of attention. The majority of the WWT procedures available today are costly, and there are diverse methods for treating minerals as well. The search for inexpensive medicines is also underway (Iakovleva and Sillanpää 2013). In this context, attention is growing in less-studied and novel MT such membrane capacitive deionization (MCDI), forward osmosis (FO), EDI, and ED. Among all the methods for WWT, the EDI process works better than equipment with a chemical basis for deionization process EDI offers flexible solutions for the chemical and energy industries. The use of this technology is beneficial to the surroundings. Future growth and attraction will keep going, especially when worker expenses and chemical prices are taken into account (Mistry et al., 2022).

5.3.2 Hydrometallurgical industry

Metal extraction and recovery from a variety of waste products have been effectively achieved using hydrometallurgical techniques. For the extraction of metals and the elimination of potentially harmful HMs present in industrial sewage, metallurgical process is a viable approach. Metal dissolving, concentration and purification, and HM recovery are the three process phases that make up the majority of hydrometallurgical techniques. Since they enable producing extremely high purities of recovered metal compounds and have a smaller effect on human well-being and the surroundings, hydrometallurgical techniques have been discovered to be a more preferred metal recycling technique than pyrometallurgical methods. As a consequence of its excellent rate of recovery, minimal energy need, lack of air pollutants, and total recycling of HM with exceptionally high purity levels, hydrometallurgy is also a widely employed recovery technique for recycling lithium-ion batteries. The hydrometallurgical technique is often carried out in stages, beginning with preprocessing and ending with leaching, segregation, and extraction of precious metals from the wastewater (Nasser and Petranikova, 2021).

The kind, level, and speciation of HM are among the chemical properties of commercial sludge that directly influence the optimal recovery strategy. While electrodeposition and precipitation are the primary techniques used in recovering metals, solvent extraction, IE, EDI, and adsorption are the key technologies utilized in concentrating and purifying (Binnemans et al., 2020). A further method used for recovering metal substances from water is electrochemical technological advances, which holds a lot of potential due to its benefits over other recuperation methods, such as its suitable affordability, outstanding versatility, ease of use, and environmentally friendly nature (Siwal et al., 2022). The main methods used for recovering metallic ions from individual ion solutions or ionic substances based on the separation characteristics of electrosorption are electrodeposition and EDI, which are electrochemical paths for metal recovery. The process of purification of solutions comprising HM ions is currently attracting a lot of interest to the EDI. However, the use of the EDI method in WWT has been significantly constrained owing to the accumulation of bivalent metal hydroxides as a consequence of metal ions interacting with hydroxide ions that are included in the EDI stack.

5.3.3 Electroplating industry

Highly hazardous cyanide, HM ions, oils and greases, organic solvents, the complex composition of effluents, as well as suspended particles, dissolved solids, total solids, and turbidity are all components of the wastewater produced by the electroplating industry (EPI). The presence of these HM ions in EPI effluent contributes to its toxicity and corrosiveness. These HM must be eliminated since they are toxic to all living things in order to keep plants, animals, and people from ingesting them. It is necessary to perform adequate WWT in order to safely release EPI effluents. For the treatment of the EPI wastewater, various methods have been developed, including flocculation and coagulation, precipitation by chemicals, IE, MT, adsorption, electrochemical treatment, and improved oxidation process (Rajoria et al., 2022).

The electrochemical cleaning procedure has a number of benefits over traditional methods, such as the full elimination of persistent organic contaminants, ecological friendliness, simplicity in integrating with other traditional methods, reduced generation of sewage, high segregation, and short residence times. The pH level, material of the electrode, operating time, electrode gap, and density of current are just a few of the variables that affect the efficacy of the electrochemical treatment procedure (Agrawal and Sahu, 2009). EDI method consists of an ED stack running concurrently with IER. The washing fluids from electroplating baths are a good example of diluted effluent that may be treated using this technique. The cost problem, which is directly connected to the salt concentration of the solutions, is one of the biggest challenges that needs to be addressed. The advancement of EDI research may lower the expenses associated with energy usage. Another significant finding is that using solutions including macromolecules increases system effectiveness, hence extending the functional life of membranes. This is especially crucial to prevent membrane blockage. In these circumstances, the creation of customized membranes and the application of EDR might be helpful techniques to extend the membrane lifetime (Scarazzato et al., 2017).

5.3.4 Pharmaceutical industries

The PI is one of the most research-intensive in the world, producing a steady stream of innovative medications that improve quality of life and save lives. Over time, the method of finding new medications has changed from being largely empirical to being heavily reliant on basic scientific understanding. Profit-driven industries and providers of fundamental research, such universities and government laboratories, have developed strong connections. Most industrialized countries have strict regulations governing the safety and effectiveness of new pharmaceutical medicines, which raises the price of clinical trials. The safeguarding of patents is exceptionally significant because to considerable costs associated with research, development, and clinical testing, as well as the potential for new goods to be easily reproduced after they have achieved success (Scherer 2000). Pharmaceutical WWT facilities often remove enough impurities from municipally treated drinking water to satisfy USP requirements for compendial waters. According to Lee et al. (2016), the phrase "compendial waters" refers to any water used to create final medication dosage kinds, comprising sterilized HPW, sterile fluid for injection, and sterile water for breathing.

HPW, sometimes referred to as HPW, has quality standards that much exceed those of water for consumption. It has been noted that HPW is often used in the production of pharmaceuticals and chemicals, semiconductors, energy production, and a number of other sectors. HPW is routinely used in laboratories to prevent problems brought on by contaminated water. A common method for producing HPW is EDI. Especially in the PI, HPW with a conductance of less than 1 μS/cm is often used in a wide range of industrial applications. Highest purity is necessary for product integrity in this business because it deals with medications. The popularity of EDI has been aided by the rising need for ultrapure water across a variety of sectors. HPW is provided by EDI, which also provides effective ion elimination and recuperation. For IER renewal, no chemicals are necessary. Strongly ionized species may be effectively and continuously removed thanks to EDI. Additionally, the product water's quality does not change over time. It is anticipated that processes that need deionized water continuously would be duplexed, with one unit supplying freshwater and the other renewing it (Saravanan et al., 2023).

5.4 Conventional techniques

Despite several technology advances and triumphs, WWT continues to be a serious problem on a global scale. If improperly processed, HM in sewage poses a serious risk to the well-being of humans, hence eliminating that it is crucial. Adsorption is the most commonly used WWT method for removing HM from sewage due to its adaptable design, simple operation, and low cost. Due to its microporous shape and simplicity of surface modification, activated carbon (AC) is the most widely used adsorbent to extract HM ions from effluent. The difficulty and expensive expense of the AC extraction from effluent solution have prevented widespread use. The recent advent of many novel substances has also demonstrated their ability to compete in eliminating the HM ions. These exceptional qualities include outstanding toughness, high chemical inertness, and an extensive surface area in these potential new materials (Chai et al., 2021). The primary traditional methods are membrane filtration (MF), electrochemical treatments, precipitation of chemicals, IE, adsorption, and coagulation—flocculation. Because of their high prices, MF and electrochemical treatments are only practical for specific technical applications (Zinicovscaia 2016).

5.4.1 Adsorption

In comparison with all other available technologies, adsorption (AD) is a straightforward, ongoing, affordable, and environmentally benign method for WWT (Rashid et al., 2021). A long-term remedy for WWT has been identified in the form of inexpensive leftovers from the residential, commercial, and farming industries. They make it possible to reduce trash, recuperate, and reuse resources while also removing toxins from effluent. AD includes the buildup of materials at the interface between two phases, such as a liquid—liquid, gas—liquid, gas—solid, or liquid—solid interface. The substance that is being adsorbed is known as the adsorbate, while the material doing the adsorbing is known as the adsorbent. The ingredients that make up adsorbates and adsorbents determine their unique characteristics. The process is referred to as physisorption if there is a physical component to the relationship between the molecules that are adsorbed and the solid surface. Because the forces that van der Waals is acting in this situation are modest, the procedure's outcomes are reversible. Additionally, it takes place below or very near the critical temperature of the adsorbate. On the contrary hand, the AD procedure is known as chemisorption if the forces of attraction among molecules that are absorbed and the surface of the material are caused by chemical bonds. Chemisorption, in contrast to physisorption, only takes place as a monolayer, and since there are greater forces at play, compounds that have been chemisorbed on solid surfaces are virtually ever removed. Both processes may take place alternately or concurrently under ideal circumstances. Physical AD is exothermic because it results in a drop in the AD system's free energy as well as entropy (De et al., 2016). Table 5.1 represents the removal of pollutants in adsorption with its operating condition.

Desorption, which is the movement of adsorbate ions from the surface of the adsorbent to the solution, frequently occurs in conjunction with AD. One may assess the potential for reversibility of AD based on the quantity of adsorbate that has been removed from the adsorbent: the more adsorbate that has been removed, the more reversible the AD process is. AD is now thought to be an effective and affordable method for eliminating harmful HM ions from wastewater effluents. This method enables the creation of treated effluents of the highest quality since it is adaptable in both construction and operation. In addition, adsorbents can be renewed by desorption since the AD is sometimes reversible. The effectiveness of adsorbents for removing HM from effluent is influenced by a variety of variables, including starting concentration, temperature, dosage of adsorbent, the pH level, interaction duration, and stirring rate. To effectively eliminate pollutants from the effluent of wastewater, materials utilized as adsorbents must have a strong sorption interaction with the intended pollutants. AC, zeolites, minerals from clay, manufacturing byproducts, waste from farming, the plant matter, and polymeric substances are a few examples of adsorbents that can be mineral, naturally occurring, or biological in nature (Burakov et al., 2018).

Although these procedures can be quite efficient in eliminating organic pollutants such as dyes, they have the drawback of creating additional waste. AD techniques are appealing methods for treating water, especially because the material that is used is inexpensive, does not need to be pretreated before use, and is simple to renew. Adsorption is an efficient and straightforward technique that removes a significant amount of dye notwithstanding its issues with adsorbent dumping and postcontamination employing adsorbents. It is also rather unconstrained in terms of producing undesirable byproducts. AD has a number of drawbacks, such as expensive adsorbent, difficult methods to separate adsorbent from dye, and small surface area. Many scientists have worked to produce inexpensive, highly effective adsorbents to handle dyes in an effort to address the drawbacks of AD (Moosavi et al., 2020).

5.4.2 Reverse osmosis

Over the past 40 years, RO MT has advanced to the point that it now accounts for 80% of DS plants deployed globally and 44% of the globe's

Table 5.1 Removal of pollutants using adsorption with its operating condition.

Contaminant	Adsorbent	Initial concentration	Adsorbent dosage	pH	Optimum time	% Removal/ adsorption capacity	References
Antimony	Zn−Fe-LDH	2 ppm	0.2 g/L	7	200 min	122.03 mg/g	Lu et al. (2015)
Arsenic	Zn−Fe-LDH	2 ppm	0.2 g/L	7	200 min	151.37 mg/g	Lu et al. (2015)
Arsenic	ZnFe-MMO	1000 μg/L	0.2 g/L	6	3 h	95.8%	Di et al. (2017)
Cadmium	Nano-alumina	50 ppm	1 g/L	5	90 min	83.33 mg/g	Afkhami et al. (2010)
Cadmium	Multiwalled carbon nanotubes	1 ppm	12.5 g/L	7	2 h	13.85%	Salam et al. (2012)
Cadmium	FAU-type zeolites	100 ppm	10 g/L	—	3 h	74.074 mg/g	Joseph et al. (2020)
Chromium	Nano-alumina	50 ppm	1 g/L	5	90 min	100 mg/g	Afkhami et al. (2010)
Cobalt	Nano-alumina	50 ppm	1 g/L	5	90 min	41.66 mg/g	Afkhami et al. (2010)
Cobalt	FAU-type zeolites	100 ppm	10 g/L	—	3 h	30.211 mg/g	Joseph et al. (2020)
Copper	Chitosan	3 ppm	0.2 g/L	5.5	95 min	20.394 mg/g	Bambaeero and Bazargan-Lari (2021)
Copper	Humic acid nano-hydroxyapatite	25 ppm	2.2 g/L	5	8 h	97.68%	Wei et al. (2020)
Copper	Multiwalled carbon nanotubes	1 ppm	12.5 g/L	7	2 h	84.98%	Salam et al. (2012)
Copper	Coconut shell-based AC	50 ppm	0.4 g/L	4	100 min	117.4 mg/g	Qin et al. (2018)
Copper	Clinoptilolite-Fe	5 ppm	10.5 g/L	8.37	48 h	99.7%	Doula (2009)
Copper	Clinoptilolite	5 ppm	10.5 g/L	7.42	48 h	99.9%	Doula (2009)
Copper	FAU-type zeolites	100 ppm	10 g/L	—	3 h	57.803 mg/g	Joseph et al. (2020)

Pollutant	Adsorbent	Concentration	Dose	pH	Time	Removal	Reference
Ibuprofen	ZnFe-MMO	5 ppm	0.5 g/L	6	3 h	95.7%	Di et al. (2017)
Lead	Nano-alumina	50 ppm	1 g/L	5	90 min	100 mg/g	Afkhami et al. (2010)
Lead	Multiwalled carbon nanotubes	1 ppm	12.5 g/L	7	2 h	31.07%	Salam et al. (2012)
Lead	FAU-type zeolites	100 ppm	10 g/L	—	3 h	103.093 mg/g	Joseph et al. (2020)
Lead	Montmorillonite-graphene oxide	100 ppm	5 g/L	6	2 h	98.94%	Zhang et al. (2019)
Manganese	Nano-alumina	50 ppm	1 g/L	5	90 min	6.29 mg/g	Afkhami et al. (2010)
Manganese	*Zoogloea* sp.	20 ppm	—	7	96 h	74.56%	Chang et al. (2021)
Manganese	Clinoptilolite-Fe	5 ppm	10.5 g/L	8.37	48 h	99.8%	Doula (2009)
Manganese	Clinoptilolite	5 ppm	10.5 g/L	7.42	48 h	53.4%	Doula (2009)
Methyl orange	Montmorillonite-pillared graphene oxide	50 ppm	4 g/L	9	2 h	97.9%	Liu et al. (2015)
Methylene Blue	Dragon fruit peel	200 ppm	0.8 g/L	10	200 min	195.2 mg/g	Jawad et al. (2021)
Methylene blue	Humic acid nano-hydroxyapatite	25 ppm	2.2 g/L	7	8 h	100%	Wei et al. (2020)
Methylene blue	Montmorillonite-pillared graphene oxide	150 ppm	0.5 g/L	7	2 h	99.9%	Liu et al. (2015)
Nickel	Nano-alumina	50 ppm	1 g/L	5	90 min	18.18 mg/g	Afkhami et al. (2010)
Nitrate	*Zoogloea* sp.	16.48 ppm	—	7	96 h	100%	Chang et al. (2021)
p-Nitrophenol	Montmorillonite-graphene oxide	150 ppm	5 g/L	6	2 h	96.82%	Zhang et al. (2019)

(*Continued*)

Table 5.1 (Continued)

Contaminant	Adsorbent	Initial concentration	Adsorbent dosage	pH	Optimum time	% Removal/ adsorption capacity	References
Tetracycline	Coconut shell-based AC	250 ppm	0.4 g/L	4	100 min	634 mg/g	Qin et al. (2018)
Tetracycline	*Zoogloea* sp.	0.4 ppm	—	7	96 h	63.59%	Chang et al. (2021)
Zinc	Chitosan	3 ppm	0.02 g/L	5.5	95 min	25.78 mg/g	Bambaeero and Bazargan-Lari (2021)
Zinc	Multiwalled carbon nanotubes	1 ppm	12.5 g/L	7	2 h	81.69%	Salam et al. (2012)
Zinc	Clinoptilolite-Fe	5 ppm	10.5 g/L	8.37	48 h	99.9%	Doula (2009)
Zinc	Clinoptilolite	5 ppm	10.5 g/L	7.42	48 h	98.7%	Doula (2009)
Zinc	FAU-type zeolites	100 ppm	10 g/L	—	3 h	42.017 mg/g	Joseph et al. (2020)

capacity currently being used for desalting. As materials have gotten better and prices have fallen, the usage of desalination membranes has expanded. RO membranes, which use specialized preprocessing and membrane layout, are now the most widely used technique for new DS sites. Seawater RO and brine water RO are two distinct forms of RO DS that have developed. The two water sources differ significantly in terms of process design, execution, and critical technical issues. These variations include foulants, salinity, alternatives for disposing of waste brine (concentration), and the location of the plant. For both kinds of RO, preprocessing choices are comparable and are based on the particulars of the water supply. RO using briny water and seawater is going to keep being utilized all over the world. Recent advances in energy recuperation and renewable energy sources, as well as creative design of plants, will enable greater utilization of DS for inland and rural regions, as well as offering cheaper water for massive coastal towns (Greenlee et al., 2009).

A membrane that is semipermeable excludes dissolved components found in the supply water in this pressure-driven procedure. Size rejection, charge being excluded, and physical—chemical relationships between the solute, solvent, and membrane are the causes of this exclusion. Performance of the procedure is influenced by operational factors, as well as membrane and feed water characteristics. Spiral-wound and hollow fiber modules are among the most often used in commerce. The last type is more fouling-prone but has an extraordinarily high density of packing and high permeates rate of production (Garud et al., 2011). The membranes might be the mixtures with multiple layers of polymers or a membrane with asymmetry with one polymer layer. The valence and strength of the membrane's charge are controlled by the functional groups that are included into the polymer's framework, and the amount of species that dissolve AD is influenced by the hydrophobicity and charge, and hardness of the membrane's surface. High area components are now accessible, lowering pressure vessel count and footprint (Malaeb and Ayoub 2011).

Due to worldwide water constraints brought on by droughts and rising populations, there is a growing interest in water reuse and recycling, which has prompted the creation of efficient WWT methods. In spite of their relatively high operating expenses and power needs, pressured MT, particularly RO, have been embraced by utilities and the WWT industry. The treatment of the concentrate has been complicated by developing pollutants, which are present in RO concentrate at greater levels than in the water used for feeding. There are presently no standards or recommendations for the evaluation or handling of newly discovered pollutants. To minimize possible effects on the well-being of humans and the surroundings, research is required addressing the handling and disposal of developing pollutants that are problematic in RO concentrate (Joo and Tansel 2015).

5.4.3 Ion-exchange process

IE occurs when two electrolytes or an electrolyte mixture and a complex exchange ions. Purification, segregation, and cleansing of aqueous or ion-containing solutions can all be accomplished through the use of IE. In order to achieve the goal of demineralization, IERs are utilized in WWT process plants (Kansara et al., 2016). IE allows for the separation of compounds or the elimination of all ions from a solution. As a result, total deionization and selective ionic pollution elimination may be differentiated. The decision between the two is mostly based on the solution's composition and the level of purification needed (Dabrowski et al., 2004). IE is a WWT method that has the benefit of being highly economical. Resin regeneration uses relatively little energy and is highly cost-effective. Resins can survive for many years before needing to be changed with proper maintenance. This specific strategy, meanwhile, comes with a lengthy list of drawbacks, making it one of the less popular ones. IER can be easily contaminated by bacterial pollutants, calcium sulphate contamination, organic material adsorption and other issues (Swanckaert et al., 2022).

5.4.4 Membrane filtration

Given its better treated water standard, effective recovery of nutrients, and economical functioning, direct MF has demonstrated tremendous promise in the WWT and resource recovery, particularly in situations where biological WWT is impractical. In these direct MF procedures, where fouling of the membrane was recognized as the primary difficulty, the parameters impacting the membrane's effectiveness as well as treatment performance are vividly highlighted. The techniques used to increase the effectiveness of direct MF, such as the preparation of feed water and the use of chemical and physical cleaning methods (Hube et al., 2020). The technique needed large flux values to be profitable at the time since membrane prices were still rather expensive. These substantial flux values were achieved using high pressures together with a variety of fouling and biofouling prevention techniques. These techniques comprised chemical and hydraulic cleaning procedures that required a lot of power, such cross-flow, backwash, and disinfecting. The cost of membranes has drastically decreased during the past 20 years. As a result, the membrane's flux has lost some of its importance in the cost estimation, and cutting back on chemical and energy consumption might result in a larger cost savings. Infrequent chemical disinfection and cleaning can minimize chemical use, while low operating pressures and efficient hydrodynamic fouling management measures may decrease the use of energy (Pronk et al., 2019).

5.4.5 Electrodialysis

Ions are moved arbitrarily across IEM in the ED process, which is a membrane separation procedure involving a field of electricity. Fig. 5.1 depicts the electrodialysis setup. Several waste or wasted solutions that are aqueous, comprising waste water from different manufacturing operations, wastewater from municipal or saline water treatment facilities, and livestock farms, have been studied to be treated by ED in both traditional and unusual ways. DS and other treatments with considerable environmental advantages may be performed with ED techniques because of their selectivity, high separating efficiency, and free of chemicals treatment. In order to reuse effluent and recover additional products, such as water, electrical power, HM ions, minerals, acids/bases, nutrients, and organic matter, ED techniques can be utilized in processes such as concentration, diluting, DS, renewal, and valorization. Intense effort has been put into creating improved or unique systems, demonstrating that zero or minimum liquid discharge techniques may be financially viable and viable from a technological standpoint (Gurreri et al., 2020).

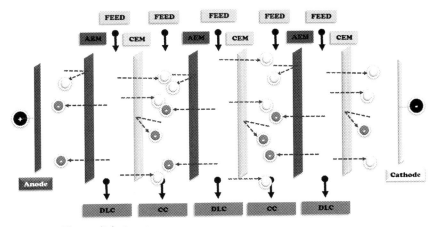

Figure 5.1 Electrodialysis setup.

5.5 Comparison of electrodeionization with other conventional techniques

According to Wang et al. (2009), AD has a number of drawbacks, including high adsorbent costs, difficulty in separating adsorbent from dye, inadequate surface area, creating waste, suitability for a small amount of treatment, ineffectiveness for dissipate dyes, requiring additional steps such as flocculation, high expenses for operation, creation of toxic byproducts, accumulation, insufficient permeability, and poor transfer of mass. Whereas with EDI, there are no trash neutralization, a lower environmental impact, more reliable operation, no servicing, and total elimination of chemicals for recuperation, over 99.9% of the ions are removed from the water when combined with RO pretreatment; it is nonpolluting, very efficient when operating big systems, safe, and reliable (Zahakifar et al., 2020).

The most efficient method for producing pure water with high pressure is RO, which makes use of a membrane that is semipermeable. Yet, the biggest problem these plants deal with is membrane biofouling, which necessitates routine membrane substitute or constant cleaning. The most popular method of disinfecting used to prior to treatment water to lessen biofouling is chlorination. Despite being widely utilized, chlorination has a number of drawbacks, including the production of byproducts that disinfect and its incapability to effectively combat a few microorganism kinds (Al-Abri et al., 2019). It is urgently necessary to produce RO membranes with interdisciplinary features and high performance with regard to both salt elimination and contamination resistance before releasing them onto the marketplace and avoiding thinking about their subsequent modification (Ghernaout and El-Wakil 2017). This is because rising levels of water pollution have outgrown the capacities of RO membranes. Polarity reversal in EDI can eliminate membrane fouling. In contrast, EDI uses less space, generates high-purity water in a steady flow, completely eliminates dissolving inorganic components, and, when combined with RO pretreatment, eliminates more than 99.9% of the ions from the water.

Since its invention in the 1940s, IE has existed in a variety of forms, including independent anion and cation beds, mixed beds, polished beds, and stacked beds, among others. IE continues to be the primary method of DI treatment. Over time, HPW systems developed with additional treatment processes created to function with IE. EDI, a different DI treatment method, first appeared in the 1980s. Initial systems had modest capacity that were useful for purifying water for laboratories, but over the years, makers of EDI equipment have created machinery that can generate the volumes of purified water required for power plants or other commercial end users. Additionally, to guarantee optimal performance, water system designers have identified the required pretreatment stages, including the usage of RO prior to EDI. The fact that EDI does not need chemical renewal is one of its advantages, which is a desirable quality. The advantage of IE is that it uses established, tested technology that, when used correctly, can provide the HPW needed by power, electronic devices, and pharmaceutical plants. Yet, chemical regeneration of the IER is necessary (Pawlowski et al., 2019).

Over the past 10 years, MF techniques involving microfiltration (MF), ultrafiltration (UF), and nanofiltration (NF) have grown quickly in the production of drinkable water. Microscopic particles and macromolecules, which typically contain inorganic particulates, organic and dissolved organic materials, are removed using MF and UF. Although MF, UF, and NF RO have outstanding results on the elimination of microbiological particles, the process of disinfection is still required; therefore, using MT does not immediately solve the problem of disinfectant byproducts. The use of membranes in surface WWT has various benefits over traditional therapy. Fouling, on the other hand, hinders this endeavor by limiting the deployment of the technology owing to the increases in hydraulic obstacles, operating and maintenance expenses, degradation of productivity, and frequently occurring membrane regenerating issues (Zularisam et al., 2006). Comparatively, EDI has a simple and continuous operation, completely removes the need for chemical renewal, has cost-effective maintenance and operation, consumes minimal power, and does not require waste neutralisation. Membrane fouling can be eliminated by polarity flipping in EDI. While RO pretreatment removes over 99.9% of the ions from the water, EDI requires less space, produces HPW in a continuous flow, totally removes dissolved inorganic components, and consumes less room overall.

5.6 Comparison of electrodeionization with electrodialysis in terms of cost and energy consumption

Electrical potential difference acts as the driving factor for ED and EDI. An electrical source creates the potential difference. A DC voltage is used in the electrochemical separation procedure known as electrodialysis (ED) to move ions via IEM. Using a driving force, the process transfers both positively and negatively charged ions from the origin water through cathodes and anodes to a concentrating effluent stream, resulting in a more diluted flow. By moving the salty water ions over a partially permeable IEM packed with an electrical potential, ED preferentially eliminates dissolved particles based on the charge they have (Valero et al., 2011). The concentrate and the dilutate, which are the cells that are produced as a result of this process, alternate between having a high concentration of ions and having an extremely low ion concentration. Water is the latter's byproduct (Huang et al., 2007). Fig. 5.2 represents the comparison of EDI with electrodialysis. In general, how EDI works is similar to how ED does. IER is present here in the DLC. Prepurified water may be processed because of the IERs strong ion transport promotion and reduction of internal impedance. A very HPW of the DLC can be attained using the IER. Gradually reaching saturation, the IER has to be renewed. Molecules of water break at the surface of the IER due to a combination of limited conductance and a high potential for electricity, the hydroxyl ions, and hydrogen ion molecules.

The IER undergoes a constant renewal process as a result of all these ions (Lee et al., 2013). Table 5.2 represents the energy consumption

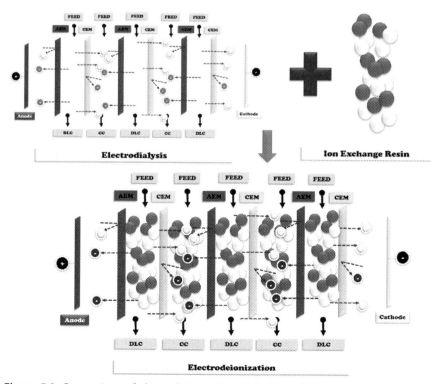

Figure 5.2 Comparison of electrodeionization with electrodialysis.

Table 5.2 Energy consumption in electrodialysis and electrodeionization (EDI) with its operating condition and device specification.

Technique	Equipment specification	Application	Operating condition	Energy consumption/ % removal	References
Electrodialysis	Compartments Ag-AgCl electrodes	Desalination	Initial concentration: 0.5 M NaCl Current density: 3 A/dm^2 Voltage: 0.3 V	150 kWh/t$_{NaCl}$	Tanaka (2003)
Electrodialysis	2 compartments 2 AEM Effective membrane area: 78.2 cm^2	Phenol removal	Initial concentration: 200 ppm Time: 120 min Voltage: 9 V	0.0075 kWh/g	Wu et al. (2019)
Electrodialysis	Multistage ED	Brackish-water	Initial concentration: 3 g/kg	134 W/m^2	Chehayeb and Nayar (2018)
Electrodialysis	Multistage ED	Desalination	Initial concentration: 35 g/kg	26 W/m^2	Chehayeb and Nayar (2018)
Electrodialysis	Multistage ED	High-salinity	Initial concentration: 70 g/kg	134 W/m^2	Chehayeb and Nayar (2018)
Electrodialysis	5 compartments 2 AEM and 2 CEM	Desalination	Initial concentration: 1 g/L Voltage: 1.4 V	0.013 kWh/m^3	Patel et al. (2021)
Electrodialysis	2-stage ED Titanium coated with Ruthenium electrode	Desalination	Initial concentration: 3.5% NaCl Flow rate: 864 mL/min Voltage: 0.2 V	0.31 kWh/kg	Yan et al. (2019)
Electrodialysis	3-stage ED	Desalination	Initial concentration: 3.5% NaCl	0.45 kWh/kg	Yan et al. (2019)

(Continued)

Table 5.2 (Continued)

Technique	Equipment specification	Application	Operating condition	Energy consumption/ % removal	References
	Titanium coated with Ruthenium electrode		Flow rate: 864 mL/min Voltage: 0.2 V		
EDI	4 compartments 1 AEM and 2 CEM	Chromium removal	Initial concentration: 100 ppm Flow rate: 12 mL/min pH: 5 Time: 6.25 h	1.21 kW h/m^3	Alvarado et al. (2009)
EDI	9 compartments Resin wafer resin	Desalination	Initial concentration: 5 g/L Flow rate: 1.1 L/min pH: 5 Time: 120 min	0.657 kWh/m^3	Pan et al. (2017)
EDI	5 compartments Resin wafer resin 2 AEM and 2 CEM	Water softening	Initial concentration: 625 ppm Flow rate: 0.6 L/min Voltage: 8.8 V Time: 70 min	0.28 kWh/m^3	Tseng et al. (2022)
EDI	9 compartments EDI Reversal 4 AEM and 4 CEM	Desalination	Initial concentration: 553 ppm Flow rate: 6 L/h Voltage: 11 V Time: 400 min	2.34 kWh/m^3	Sun et al. (2016)

AEM, Anionic exchange membrane; *CEM*, cationic exchange membrane; *ED*, electrodialysis.

in electrodialysis and EDI with its operating condition and device specification.

By introducing IER, or ionically conducting medium, into the CC, EDI overcomes this difficulty. As a result, the ions may readily exit the CC and DLC without the need for high voltage. However, the sluggish rate at which ions flow through water restricts this process. In practice, ion removal is hampered by water's poor conductance. In other words, the electrical resistivity of the water in the DLC rises as the water gets higher in purity. By placing a combination of CER and AER between the two membranes, EDI technology addresses this issue. The IER beads' vast surface area now effectively removes the ion diffusion resistance, allowing the ions to travel easily. The resin bead surface functions as a conducting channel even in HPW solution, effectively functioning as an interface for the ions to migrate toward the membrane surface more quickly than they would have in water alone. As a result, EDI is more environmentally friendly than ED because to its low use of energy, lack of chemical renewal, chemical disposal, lack of resin waste, and low operating costs (Kumar and Pan 2020).

5.7 Conclusion

The goal of EDI, a technique that was developed 60 years ago, is to reduce the concentration polarization phenomena that exists in ED systems. Due to its ability to remove a variety of impurities, EDI offers flexible solutions to the separations in sectors including biochemical, chemical, food, MI, and PI. This chapter looked at EDI and its applications in the mining, hydrometallurgical, electroplating, and pharmaceutical industries, among others. Additionally, it examined other conventional techniques including AD, RO, IE process, MF, and ED and contrasted them with EDI in terms of costs and energy consumption. The simple and continuous functioning of EDI completely removes the need for chemical renewal, offers affordable service and operation, requires minimal power, and does not call for waste neutralization. Polarity flipping in EDI helps get rid of membrane fouling. While RO pretreatment eliminates more than 99.9% of the ions from the water, EDI uses up less space, generates HPW in a continuous flow, completely eliminates dissolved inorganic components, and takes up less space overall. As a result of its low energy consumption, the absence of chemical renewal, lack of chemical disposal, lack of resin waste, and low running expenses, EDI is therefore more ecologically friendly than ED.

References

Afkhami, A., Saber-Tehrani, M., Bagheri, H., 2010. Simultaneous removal of heavy-metal ions in wastewater samples using nano-alumina modified with 2, 4-dinitrophenylhydrazine. Journal of Hazardous Materials 181 (1–3), 836–844.

Agrawal, A., Sahu, K.K., 2009. An overview of the recovery of acid from spent acidic solutions from steel and electroplating industries. Journal of Hazardous Materials 171 (1–3), 61–75.

Al-Abri, M., Al-Ghafri, B., Bora, T., Dobretsov, S., Dutta, J., Castelletto, S., et al., 2019. Chlorination disadvantages and alternative routes for biofouling control in reverse osmosis desalination. npj Clean Water 2 (1). Available from: https://doi.org/10.1038/s41545-018-0024-8, https://www.nature.com/npjcleanwater/.

Alvarado, L., Chen, A., 2014. Electrodeionization: principles, strategies and applications. Electrochimica Acta 132, 583–597.

Alvarado, L., Ramírez, A., Rodríguez-Torres, I., 2009. Cr (VI) removal by continuous electrodeionization: study of its basic technologies. Desalination 249 (1), 423–428.

Arar, O., Yüksel, U., Kabay, N., Yüksel, M., 2014. Various applications of electrodeionization (EDI) method for water treatment—a short review. Desalination 342, 16–22. Available from: https://doi.org/10.1016/j.desal.2014.01.028.

Bambaeero, A., Bazargan-Lari, R., 2021. Simultaneous removal of copper and zinc ions by low cost natural snail shell/hydroxyapatite/chitosan composite. Chinese Journal of Chemical Engineering 33, 221–230.

Binnemans, K., Jones, P.T., Manjón Fernández, Á., Masaguer Torres, V., 2020. Hydrometallurgical processes for the recovery of metals from steel industry by-products: a critical review. Journal of Sustainable Metallurgy 6 (4), 505–540. Available from: https://doi.org/10.1007/s40831-020-00306-2, https://link.springer.com/journal/40831.

Burakov, A.E., Galunin, E.V., Burakova, I.V., Kucherova, A.E., Agarwal, S., Tkachev, A.G., et al., 2018. Adsorption of heavy metals on conventional and nanostructured materials for wastewater treatment purposes: a review. Ecotoxicology and Environmental Safety 148, 702–712. Available from: https://doi.org/10.1016/j.ecoenv.2017.11.034, http://www.elsevier.com/inca/publications/store/6/2/2/8/1/9/index.htt.

Carvalho, F.P., 2017. Mining industry and sustainable development: time for change. Food and Energy security 6 (2), 61–77.

Chai, W.S., Cheun, J.Y., Kumar, P.S., Mubashir, M., Majeed, Z., Banat, F., et al., 2021. A review on conventional and novel materials towards heavy metal adsorption in wastewater treatment application. Journal of Cleaner Production 296. Available from: https://doi.org/10.1016/j.jclepro.2021.126589, https://www.journals.elsevier.com/journal-of-cleaner-production.

Chang, Q., Ali, A., Su, J., Wen, Q., Bai, Y., Gao, Z., 2021. Simultaneous removal of nitrate, manganese, and tetracycline by *Zoogloea* sp. MFQ7: adsorption mechanism of tetracycline by biological precipitation. Bioresource Technology 340. Available from: https://doi.org/10.1016/j.biortech.2021.125690.

Chehayeb, K.M., Nayar, K.G., 2018. On the merits of using multi-stage and counterflow electrodialysis for reduced energy consumption. Desalination 439, 1–16.

Dąbrowski, A., Hubicki, Z., Podkościelny, P., Robens, E., 2004. Selective removal of the heavy metal ions from waters and industrial wastewaters by ion-exchange method. Chemosphere 56 (2), 91−106.

Damigos, D., 2006. An overview of environmental valuation methods for the mining industry. Journal of Cleaner Production 14 (3−4), 234−247.

De, Gisi, S., Lofrano, G., Grassi, M., Notarnicola, M., 2016. Characteristics and adsorption capacities of low-cost sorbents for wastewater treatment: a review. Sustainable Materials and Technologies 9, 10−40.

Di, G., Zhu, Z., Zhang, H., Zhu, J., Lu, H., Zhang, W., et al., 2017. Simultaneous removal of several pharmaceuticals and arsenic on Zn-Fe mixed metal oxides: combination of photocatalysis and adsorption. Chemical Engineering Journal 328, 141−151. Available from: https://doi.org/10.1016/j.cej.2017.06.112, http://www.elsevier.com/inca/publications/store/6/0/1/2/7/3/index.htt.

Doula, M.K., 2009. Simultaneous removal of Cu, Mn and Zn from drinking water with the use of clinoptilolite and its Fe-modified form. Water Research 43 (15), 3659−3672.

Garud, R.M., Kore, S.V., Kore, V.S., Kulkarni, G.S., 2011. A short review on process and applications of reverse osmosis. Universal Journal of Environmental Research & Technology (3), .

Ghernaout, D., El-Wakil, A., 2017. Requiring reverse osmosis membranes modifications—an overview. American Journal of Chemical Engineering 5 (4), 81−88.

Greenlee, L.F., Lawler, D.F., Freeman, B.D., Marrot, B., Moulin, P., 2009. Reverse osmosis desalination: water sources, technology, and today's challenges. Water Research 43 (9), 2317−2348. Available from: https://doi.org/10.1016/j.watres.2009.03.010, http://www.elsevier.com/locate/watres.

Gurreri, L., Tamburini, A., Cipollina, A., Micale, G., 2020. Electrodialysis applications in wastewater treatment for environmental protection and resources recovery: a systematic review on progress and perspectives. Membranes 10 (7). Available from: https://doi.org/10.3390/membranes10070146.

Hakim, A.N., Khoiruddin, K., Ariono, D., Wenten, I.G., 2020. Ionic separation in electrodeionization system: mass transfer mechanism and factor affecting separation performance. Separation & Purification Reviews 49 (4), 294−316. Available from: https://doi.org/10.1080/15422119.2019.1608562.

Huang, C., Xu, T., Zhang, Y., Xue, Y., Chen, G., 2007. Application of electrodialysis to the production of organic acids: state-of-the-art and recent developments. Journal of Membrane Science 288 (1−2), 1−12. Available from: https://doi.org/10.1016/j.memsci.2006.11.026.

Hube, S., Eskafi, M., Hrafnkelsdóttir, K.F., Bjarnadóttir, B., Bjarnadóttir, M.Á., Axelsdóttir, S., et al., 2020. Direct membrane filtration for wastewater treatment and resource recovery: a review. Science of the Total Environment 710. Available from: https://doi.org/10.1016/j.scitotenv.2019.136375, http://www.elsevier.com/locate/scitotenv.

Iakovleva, E., Sillanpää, M., 2013. The use of low-cost adsorbents for wastewater purification in mining industries. Environmental Science and Pollution Research 20, 7878−7899.

Jawad, A.H., Saud Abdulhameed, A., Wilson, L.D., Syed-Hassan, S.S.A., ALOthman, Z.A., Khan, M.R., 2021. High surface area and mesoporous activated carbon from KOH-activated dragon fruit peels for methylene blue dye adsorption: optimization and mechanism study. Chinese Journal of Chemical Engineering 32, 281−290. Available from: https://doi.org/10.1016/j.cjche.2020.09.070, https://www.journals.elsevier.com/chinese-journal-of-chemical-engineering.

Joo, S.H., Tansel, B., 2015. Novel technologies for reverse osmosis concentrate treatment: a review. Journal of Environmental Management 150, 322−335.

Joseph, I.V., Tosheva, L., Doyle, A.M., 2020. Simultaneous removal of Cd (II), Co (II), Cu (II), Pb (II), and Zn (II) ions from aqueous solutions via adsorption on FAU-type zeolites prepared from coal fly ash. Journal of Environmental Chemical Engineering 8 (4), 103895.

Kansara, N., Bhati, L., Narang, M., Vaishnavi, R., 2016. Wastewater treatment by ion exchange method: a review of past and recent researches. ESAIJ (Environmental Science 12 (4), 143−150.

Kumar P.S., Varsha M., Rathi B.S., Rangasamy G., 2022. Electrodeionization: Fundamentals, Methods and Applications. Environmental Research.

Kumar, A., Pan, S.Y., 2020. Opportunities and challenges of electrochemical water treatment integrated with renewable energy at the water-energy nexus. Water-Energy Nexus 3, 110−116.

Lee, H.J., Song, J.H., Moon, S.H., 2013. Comparison of electrodialysis reversal (EDR) and electrodeionization reversal (EDIR) for water softening. Desalination 314, 43−49.

Lee, H., Jin, Y., Hong, S., 2016. Recent transitions in ultrapure water (UPW) technology: rising role of reverse osmosis (RO). Desalination 399, 185−197.

Liu, L., Zhang, B., Zhang, Y., He, Y., Huang, L., Tan, S., et al., 2015. Simultaneous removal of cationic and anionic dyes from environmental water using montmorillonite-pillared graphene oxide. Journal of Chemical and Engineering Data 60 (5), 1270−1278. Available from: https://doi.org/10.1021/je5009312, http://pubs.acs.org/journal/jceaax.

Lu, H., Zhu, Z., Zhang, H., Zhu, J., Qiu, Y., 2015. Simultaneous removal of arsenate and antimonate in simulated and practical water samples by adsorption onto Zn/Fe layered double hydroxide. Chemical Engineering Journal 276, 365−375. Available from: https://doi.org/10.1016/j.cej.2015.04.095, http://www.elsevier.com/inca/publications/store/6/0/1/2/7/3/index.htt.

Malaeb, L., Ayoub, G.M., 2011. Reverse osmosis technology for water treatment: state of the art review. Desalination 267 (1), 1−8.

Mistry, G., Popat, K., Patel, J., Panchal, K., Ngo, H.H., Bilal, M., et al., 2022. New outlook on hazardous pollutants in the wastewater environment: occurrence, risk assessment and elimination by electrodeionization technologies. Environmental Research 115112.

Moosavi, S., Lai, C.W., Gan, S., Zamiri, G., Akbarzadeh Pivehzhani, O., Johan, M.R., 2020. Application of efficient magnetic particles and activated carbon for dye removal from wastewater. ACS Omega 5 (33), 20684−20697. Available from: https://doi.org/10.1021/acsomega.0c01905, pubs.acs.org/journal/acsodf.

Nasser, O.A., Petranikova, M., 2021. Review of achieved purities after li-ion batteries hydrometallurgical treatment and impurities effects on the cathode performance. Batteries 7 (3), 60.

Otero, C., Urbina, A., Rivero, E.P., Rodríguez, F.A., 2021. Desalination of brackish water by electrodeionization: experimental study and mathematical modeling. Desalination 504. Available from: https://doi.org/10.1016/j.desal.2020.114803.

Pan, S.Y., Snyder, S.W., Ma, H.W., Lin, Y.J., Chiang, P.C., 2017. Development of a resin wafer electrodeionization process for impaired water desalination with high energy efficiency and productivity. ACS Sustainable Chemistry and Engineering 5 (4), 2942−2948. Available from: https://doi.org/10.1021/acssuschemeng.6b02455, http://pubs.acs.org/journal/ascecg.

Patel, S.K., Biesheuvel, P.M., Elimelech, M., 2021. Energy consumption of brackish water desalination: identifying the sweet spots for electrodialysis and reverse osmosis. ACS ES&T Engineering 1 (5), 851−864.

Pawlowski, S., Crespo, J.G., Velizarov, S., 2019. Profiled ion exchange membranes: a comprehensible review. International Journal of Molecular Sciences 20 (1), 165.

Peng, H., Guo, J., 2020. Removal of chromium from wastewater by membrane filtration, chemical precipitation, ion exchange, adsorption electrocoagulation, electrochemical reduction, electrodialysis, electrodeionization, photocatalysis and nanotechnology: a review. Environmental Chemistry Letters 18, 2055—2068.

Pronk, W., Ding, A., Morgenroth, E., Derlon, N., Desmond, P., Burkhardt, M., et al., 2019. Gravity-driven membrane filtration for water and wastewater treatment: a review. Water Research 149, 553—565. Available from: https://doi.org/10.1016/j.watres.2018.11.062, http://www.elsevier.com/locate/watres.

Qin, Q., Wu, X., Chen, L., Jiang, Z., Xu, Y., 2018. Simultaneous removal of tetracycline and Cu(II) by adsorption and coadsorption using oxidized activated carbon. RSC Advances 8 (4), 1744—1752. Available from: https://doi.org/10.1039/c7ra12402c, http://pubs.rsc.org/en/journals/journal/ra.

Rajoria, S., Vashishtha, M., Sangal, V.K., 2022. Treatment of electroplating industry wastewater: a review on the various techniques. Environmental Science and Pollution Research 29 (48), 72196—72246.

Rashid, R., Shafiq, I., Akhter, P., Iqbal, M.J., Hussain, M., 2021. A state-of-the-art review on wastewater treatment techniques: the effectiveness of adsorption method. Environmental Science and Pollution Research 28 (8), 9050—9066. Available from: https://doi.org/10.1007/s11356-021-12395-x, https://link.springer.com/journal/11356.

Rathi, B.S., Kumar, P.S., 2020. Electrodeionization theory, mechanism and environmental applications. A review. Environmental Chemistry Letters 18 (4), 1209—1227. Available from: https://doi.org/10.1007/s10311-020-01006-9, http://springerlink.metapress.com/app/home/journal.asp?wasp = d86tgdwvtg0yvw9gvkwp&referrer = parent&backto = browsepublicationsresults,140,541.

Rathi, B.S., Kumar, P.S., Parthiban, R., 2022. A review on recent advances in electrodeionization for various environmental applications. Chemosphere 289, 133223.

Razzak, S.A., Faruque, M.O., Alsheikh, Z., Alsheikhmohamad, L., Alkuroud, D., Alfayez, A., et al., 2022. A comprehensive review on conventional and biological-driven heavy metals removal from industrial wastewater. Environmental Advances 7, 100168.

Salam, M.A., Al-Zhrani, G., Kosa, S.A., 2012. Simultaneous removal of copper (II), lead (II), zinc (II) and cadmium (II) from aqueous solutions by multi-walled carbon nanotubes. Comptes rendus chimie 15 (5), 398—408.

Saravanan, A., Yaashikaa, P.R., Senthil Kumar, P., Karishma, S., Thamarai, P., Deivayanai, V.C., Rangasamy, G., Selvasembian, R., Aminabhavi, T.M., 2023. Environmental sustainability of toxic arsenic ions removal from wastewater using electrodeionization. Separation and Purification Technology 317. Available from: https://doi.org/10.1016/j.seppur.2023.123897.

Scarazzato, T., Panossian, Z., Tenório, J.A.S., Pérez-Herranz, V., Espinosa, D.C.R., 2017. A review of cleaner production in electroplating industries using electrodialysis. Journal of Cleaner Production 168, 1590—1602. Available from: https://doi.org/10.1016/j.jclepro.2017.03.152.

Scherer, F.M., 2000. The pharmaceutical industry. Handbook of Health Economics 1, 1297—1336.

Shrestha, R., Ban, S., Devkota, S., Sharma, S., Joshi, R., Tiwari, A.P., et al., 2021. Technological trends in heavy metals removal from industrial wastewater: a review. Journal of Environmental Chemical Engineering 9 (4). Available from: https://doi.org/10.1016/j.jece.2021.105688, http://www.journals.elsevier.com/journal-of-environmental-chemical-engineering/.

Siwal, S.S., Kaur, H., Deng, R., Zhang, Q., 2022. A review on electrochemical techniques for metal recovery from waste resources. Current Opinion in Green and Sustainable Chemistry 100722.

Sun, X., Lu, H., Wang, J., 2016. Brackish water desalination using electrodeionization reversal. Chemical Engineering and Processing: Process Intensification 104, 262−270.

Swanckaert, B., Geltmeyer, J., Rabaey, K., De Buysser, K., Bonin, L., De Clerck, K., 2022. A review on ion-exchange nanofiber membranes: properties, structure and application in electrochemical (waste)water treatment. Separation and Purification Technology 287. Available from: https://doi.org/10.1016/j.seppur.2022.120529.

Tanaka, Y., 2003. Mass transport and energy consumption in ion-exchange membrane electrodialysis of seawater. Journal of Membrane Science 215 (1−2), 265−279.

Tate, J., 2014. Comparison of continuous electrodeionization technologies. IWC 14, 01.

Tseng, P.C., Lin, Z.Z., Chen, T.L., Lin, Y., Chiang, P.C., 2022. Performance evaluation of resin wafer electrodeionization for cooling tower blowdown water reclamation. Sustainable Environment Research 32 (1). Available from: https://doi.org/10.1186/s42834-022-00145-8, https://link.springer.com/journal/42834.

Valero, F., Barceló, A., Arbós, R., 2011. Electrodialysis technology: theory and applications. Desalination, Trends and Technologies 28, 3−20.

Wang, L.W., Wang, R.Z., Oliveira, R.G., 2009. A review on adsorption working pairs for refrigeration. Renewable and Sustainable Energy Reviews 13 (3), 518−534.

Wei, W., Han, X., Zhang, M., Zhang, Y., Zheng, C., 2020. Macromolecular humic acid modified nano-hydroxyapatite for simultaneous removal of Cu(II) and methylene blue from aqueous solution: experimental design and adsorption study. International Journal of Biological Macromolecules 215, 849−860. Available from: https://doi.org/10.1016/j.ijbiomac.2020.02.137, http://www.elsevier.com/locate/ijbiomac.

Wu, D., Chen, G.Q., Hu, B., Deng, H., 2019. Feasibility and energy consumption analysis of phenol removal from salty wastewater by electro-electrodialysis. Separation and Purification Technology 215, 44−50. Available from: https://doi.org/10.1016/j.seppur.2019.01.001, http://www.journals.elsevier.com/separation-and-purification-technology/.

Yan, H., Wang, Y., Wu, L., Shehzad, M.A., Jiang, C., Fu, R., et al., 2019. Multistage-batch electrodialysis to concentrate high-salinity solutions: process optimisation, water transport, and energy consumption. Journal of Membrane Science 570−571, 245−257. Available from: https://doi.org/10.1016/j.memsci.2018.10.008, http://www.elsevier.com/locate/memsci.

Zahakifar, F., Keshtkar, A.R., Souderjani, E.Z., Moosavian, M.A., 2020. Use of response surface methodology for optimization of thorium(IV) removal from aqueous solutions by electrodeionization (EDI). Progress in Nuclear Energy. Available from: https://doi.org/10.1016/j.pnucene.2020.103335.

Zhang, C., Luan, J., Yu, X., Chen, W., 2019. Characterization and adsorption performance of graphene oxide − montmorillonite nanocomposite for the simultaneous removal of Pb^{2+} and p-nitrophenol. Journal of Hazardous Materials 378, 120739. Available from: https://doi.org/10.1016/j.jhazmat.2019.06.016.

Zinicovscaia, I., 2016. Conventional methods of wastewater treatment. Cyanobacteria for Bioremediation of Wastewaters. pp. 17−25.

Zularisam, A.W., Ismail, A.F., Salim, R., 2006. Behaviours of natural organic matter in membrane filtration for surface water treatment—a review. Desalination 194 (1−3), 211−231.

CHAPTER SIX

Heavy metal ions removal by electrodeionization

6.1 Introduction

Heavy metals (HMs) are naturally occurring substances with large atomic weights and mass-to-weight ratios, which are a minimum of five times denser than the density of water. Their broad distribution in the atmosphere as an outcome of their multiple industrial, households agricultural, therapeutic, and technological uses has raised concerns about the potential impact on both human well-being and the ecology. How dangerous it is depending on the quantity consumed, the method of interaction, the substance, and also the gender, age, transmission, and nutritional status of persons who were subjected. Because of their extreme amount of toxic exposure, cadmium, arsenic, lead, chromium, and mercury are some of the most dangerous metals that the general public is concerned about. These metals are recognized as systemically hazardous chemicals since they have been associated with damaging organs in different ways even at low exposure levels. Both the US Environmental Protection Agency and the International Agency for Research on Cancer list these substances as probable human carcinogens (Tchounwou et al., 2012). In order to avoid contaminating the surroundings, which is fundamentally vital to life, it is also crucial to understand their origins, leach techniques, chemical transformations, and patterns of accumulation. They either dissolve into underground water where they travel through water pathways, then accumulate in the groundwater, or they flow off into the earth's surface where they pollute both the drinking water and the land. Animal transmission and collective processes commonly result in toxicity and damage (Duruibe et al., 2007).

HM ions are toxic to people and aquatic organisms in aqueous systems when their concentration exceeds safe limits. It is impossible for HMs to decay. Instead, they may enter directly or via the food chain and then build up in creatures that live. Metal ions can either be directly harmful to the metabolism or transformed inside the human body into more poisonous forms (Malik et al., 2019). Over the last few decades, a wide range of methods for eliminating HM from effluent have been widely investigated. These methods include ion exchange, electrochemical processes, adsorption, membrane filtration, chemical precipitation, coagulation–flocculation, and flotation. The review of published articles makes it clear that membrane filtration, ion exchange, and adsorption are the processes that are most commonly researched for the cleanup of HM in effluent (Fu and Wang, 2011).

The creation of sustainable pollution-removal technology is required due to the increasing poisoning of environments. In particular, electrodeionization (EDI) has recently been shown to be a successful technique for eliminating ionic compounds from contaminated rivers (Mistry et al., 2022). The largest factor in EDI succeeds is undoubtedly the reality that it is a cleaner process. It does deal with the wastewater associated with substitution for resin rejuvenation electrical energy for the harmful substances that were previously employed to replenish resins. In the treatment of water, EDI eliminates some of the problems associated with ion–exchange resin (IER) beds, most notably ion dumping when the beds wear out (Rathi et al., 2022). Benefits include the elimination of corrosive anions, HM ions, radioactive contaminants, and hazardous substances, as well as the recovery of precious metals, the creation of large amounts of purified water, and the nitrate elimination. Optimizing efficiency and changing the stack arrangement are required for EDI (Kumar et al., 2022).

As of now, EDI has excelled as a very effective, ecologically friendly process of removal, separation, and concentrating. New materials are constantly being created to advance and mature this technology, which might have huge positive effects on our surroundings and the global economy (Arar et al., 2014). The aim of this chapter was to give an extensive summary of HMs, their causes, and their effects. It has also assessed several methods of EDI-based HM removal. With the help of the offered applications examples, it has been shown that the EDI technique is quite efficient in recovering some important species, such HMs. Due to EDI, a very high percentage of these pollutants are consistently eliminated.

6.2 Heavy metals

HMs are a natural part of the environment, having both necessary and unnecessary forms. HM toxicity refers to an excessive of the necessary amount or an undesirable concentration that has been identified instinctively in the earth and became concentrated as a consequence of human-caused actions. HMs may penetrate plant, animal, and human organs through breathing in, eating habits, and handling by hand. They may attach to and impair the operation of essential cellular components. These are a collection of metals and metalloids, which also include platinum, chromium, copper, arsenic, silver, lead, cadmium, nickel, cobalt, zinc, and iron (Asati et al., 2016). The most significant risks to human health associated with being exposed to HM include As, Hg, pb, and Cd. International organizations such as the WHO constantly assess the impact of these HM have on human wellness after doing considerable research on them. HM has been used by human beings for quite a while. Although the reality that over decades, pollution levels have decreased in a large number of affluent countries, HM being exposed continues to rise and in certain locations is even increasing, particularly in less developed countries. This is despite the reality that some of the adverse health consequences of HM are well known (Järup, 2003).

6.2.1 Sources

Natural as well as synthetic sources of HM in the environment are possible. Although HM poisoning of the surroundings has long been recognized as a serious issue, it has recently gotten worse and there are few practical solutions (He et al., 2013). Human influences include mineral extraction, abandoned mining zones, industrialization, urbanization, the use of fertilizer and pesticides in agricultural activities, domestic sewage, and irrigation purposes of effluent. Natural causes comprise the movement and rearrangement of rock weathering, soil litter, volcanic eruption, erosion, and the hydraulic movement of soil-derived rock with a significant background value under the influence of wind (Zhang and Wang, 2020). The many causes of HM are shown in Fig. 6.1.

Through the soil, HMs from fertilizers containing phosphates, urban organic matter, and sewage waste are carried to aquifers. HM may be included in these kinds of chemical fertilizers, which are mostly composed of phosphorus and nitrogen components. Additionally, the use of fertilizers and pesticides in current cultivation methods to safeguard and increase yields of crops has resulted in the long-term persistence of their leftovers in farming soils, which are then carried to aquatic environments via rainfall or water from irrigation. Additionally, foods cultivated in HM-

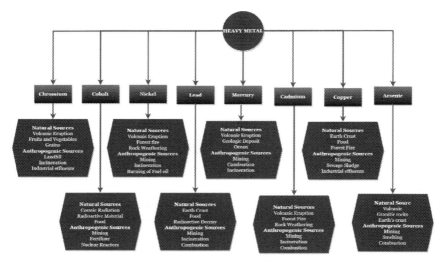

Figure 6.1 Various sources of HM. *HM*, Heavy metal.

contaminated soils are fed to animals whose feces is then utilized in the aquaculture industry passing the metals into the environment through sediment in the process (Emenike et al., 2021).

Some elements have a relationship with the elements from nature like soil origin or parent material, notably the levels of chromium and manganese. The naturally occurring level of HMs in the soil varies greatly due to the variability of kinds of soil and parent substances. Another factor contributing to the large build-up of HMs is the acidity of the soils. The levels of arsenic in the natural earth's crust are significant; however, natural sources are not the primary causes of HM poisoning of soil. Therefore, human activities account for the majority of the causes of HM component emission into land (Sodango et al., 2018). Significant HMs are released into the environment, water, and land, including Pb, As, Hg, Cr, and Cd. They are not biodegradable, so HMs have the potential to biomagnify and enter into the food chain. As a result, the enhanced HM can now enter the human body. Even while certain HMs, including cobalt, copper, zinc, and a few other vital minor elements, are required for life's essential functions, all metals can be fatal in excess amounts (Hu et al., 2014).

One or more HMs accumulate in the soil as a result of human factors disrupting the naturally occurring, gradually happening geochemical cycle of HMs. The dose of HMs in the farmland environment is increased by human-caused processes such the usage of chemicals used for pest, disposal of urban garbage, the combustion of petroleum and coal for power, the extraction of minerals and their smelting processes, the watering of fields with wastewater, and the usage of fertilizer. The industries that contribute to the presence of HMs in river water include those that make metals, paints, pigments, galvanizing iron products, pulp and paper, varnishes, rubber, rayon, thermal power plants, cotton textiles, steel plants, mining, and unsystematic use of pesticides and fertilizers that contain HMs in fields for farming. These HMs have a built-up impact in aquifers and water for drinking at low concentrations (Paul, 2017; Li et al., 2019).

6.2.2 Effects of heavy metals

HM contamination in the atmosphere is becoming progressively problematic, and it is raising widespread concern owing to the negative consequences. Such dangerous compounds are being released into our waterways, land, and ecosystem as a result of the fast growing farming and metal industries, ineffective disposal of waste, synthetic fertilizers, and pesticides (Briffa et al., 2020). Due to its poisonous nature, HM poisoning poses substantial health concerns for humans hazards and our surroundings (Jacob et al., 2018). The adverse effects of HM on humans are shown in Fig. 6.2.

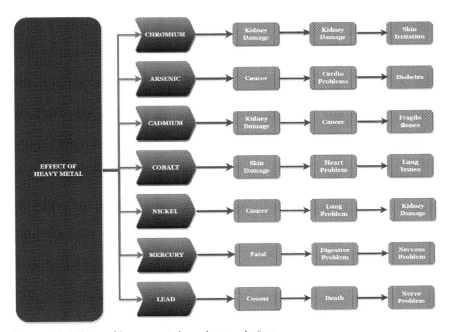

Figure 6.2 Effects of heavy metals on human beings.

Trivalent chromium is a minor element that is crucial for both humans and animals. Pure Cr has no negative effects. When present in extremely high concentrations, trivalent chromium is thought to have little harmful impact. Hexavalent compounds are mostly to blame for the chromium toxic effects, both acute and long-term. Allergic and eczematous skin reactions, dermatitis, renal oligo anuric deficiency, perforation of the allergic suffering from asthma responses, epidermal and mucus membrane ulcerations, bronchial cancers known as carcinoma and the septum of the nose, hepatocellular deficiency, and gastro-enteritis are the most significant adverse reactions after interaction, breathing in, or consumption of hexavalent chromium chemicals (Baruthio, 1992). Chromium penetrates the human body via the skin, lungs, and digestive system, with some skin absorption as well. In contrast to nonoccupational exposure, which happens when people consume chromium-containing foods and beverages, breathing is the most significant route for exposure while working. Cr(III) has a poor absorption regardless of the exposure method, but Cr(VI) is more easily absorbed. Additionally, because Cr(VI) absorbs less when taken orally, it is not very harmful when taken that way. However, chromium is extremely hazardous and induces lung tumors, irritation of the nose, nasal ulcers, hypersensitive responses, and allergic reactions when ingested orally or topically. Before reaching the blood stream, all of the consumed Cr(VI) is converted to Cr(III) in the body (Shrivastava et al., 2002).

A typical ecological pollutant that is found all around the globe is arsenic (As). Water from wells and polluted soil, fish as well as other marine organisms rich in methylated arsenic species, working environments, and other factors all contribute to the exposure of humans to this metalloid. Human contact with arsenic is linked to a number of health issues, including malignancy, damage to the liver, dermatosis, and disruptions of the neurological system such polyneuropathy, EEG deviations, and, in severe cases, illusions, confusion, and anxiety (Rodríguez et al., 2003; Bjørklund et al., 2018). In numerous areas of the cerebral cortex, including the pituitary gland, arsenic accumulates in various forms, including inorganic and methylated arsenicals. Arsenic is also a recognized teratogen that penetrates the membrane of the placenta when a fetus is developing and, in high quantities, causes growth problems and defects in the neural tube (Tyler and Allan, 2014). Though arsenic has been identified as a diabetes type 2 associated risk factor with a particular system, it is still unknown whether arsenic also acts as an obesogen. Different research have examined the impact of arsenic on white adipose tissue (WAT), but only in certain of them for fat-related goals. The disruption of WAT metabolism is critical in the genesis of obese (Ceja-Galicia et al., 2017; Paul et al., 2007).

In the natural world, cobalt and its compounds are widely distributed and utilized in a wide range of human activities. As a metallic component of vitamin B12, cobalt performs a medically necessary role, but prolonged exposure has been related to a range of adverse health effects. The effects on overall wellness are described as a complex clinical disorder that primarily involves neuro (including hearing and eye problems), cardiovascular, and endocrine abnormalities (Leyssens et al., 2017). Cobalt alloys or compounds have been linked to the indications of potential cancer-causing effects in human beings due to its usage in medicine, the hard metals industry, and cobalt manufacturing (Jensen and Tuchsen, 1990).

Nickel, a transition component, circulates widely in surroundings, particularly soil, water, and the air. Despite Ni is a widespread element in the environment, its value as an essential component for humans as well as animals has not yet been recognized. Industries, the consumption of fuels, in addition to waste from municipalities and factories may all contribute to nickel contamination in the atmosphere. People who come into touching nickel may have a variety of harmful health effects, such as allergy symptoms, cardiac, kidney, and lung inflammation, as well as bronchial and nasal cancer. It has been suggested that mitochondrial dysfunction and oxidative stress serve a basic and vital role in the toxicity caused by nickel, despite the fact that the precise molecular pathways generating this metal's toxic effects are yet unknown (Genchi et al., 2020). The rise of nickel-derived allergens, nickel-induced tumor development, and infectious illnesses brought on by pathogenic organisms that depend on nickel-based digestive enzymes to colonize humans (Zambelli and Ciurli, 2013; Zambelli et al., 2016).

The exposure to lead is one of the most hazardous metal toxicities. It may enter the body of an individual through paint containing lead, debris, soil, water, cutlery, and folk treatments. Children are more likely to become poisoned with lead. Because there is not enough glutamate substitution, lead damages cells by inducing a process called oxidative stress. Pb may additionally trigger serious anemia as a consequence of the membrane of cells being destroyed by peroxidation of lipids (Debnath et al., 2019; Demayo et al., 1982). Lead (Pb) is among the determined, common HM that has been linked to a variety of harmful consequences in animals. The testicles of both humans and wildlife may change in how they function as a result of exposure to this metal. The health of the public is seriously threatened by lead poisoning, particularly in underdeveloped nations. As a result, significant efforts have been made on the part of workplace and public health to reduce the risks associated with this metal. The hematological, reproductive, renal, and brain systems of human beings are among those that are particularly vulnerable to dangers following interaction in high Pb concentrations (Assi et al., 2016).

Worldwide concern surrounds mercury contamination and its effects on the well-being of people (Ha et al., 2017). The bulk of human contact with mercury and the accompanying health concerns are brought on by eating seafood and aquatic animals (Sundseth et al., 2017). Since it has been linked to Minamata illness, methylmercury, which is produced when bivalent mercury is broken down, is a well-known neurotoxin. It bioadsorbs into the atmosphere and biomagnifies throughout the food chain. Fish and seafood eating accounts for the majority of human exposure to methylmercury. The brain's nervous system is damaged by methylmercury, which readily crosses the blood—brain barrier, especially in fetuses (Honda et al., 2006). Elderly kidneys, who may not be performing at their peak, might be more susceptible than normal kidneys to the adverse impacts of Hg since normal kidneys may suffer glomerular and tubular harm when exposed to mercuric ions. The consequences of Hg on the kidney have been distinguished from related to age kidney abnormalities. On the other hand, little is known about how nephrotoxicants, such as mercury, affect the kidneys as they age (Bridges and Zalups, 2017).

Cd is successfully retained in the human body after ingestion, where it accumulates during one's entire life. The principal location of accumulation for Cd, the proximal tubular cells, is harmful to the kidney. Additionally, cadmium can demineralize bones either straight via bone destruction or indirectly via renal failure. In the workplace, prolonged exposure to airborne Cd can harm the lungs and raise the chance of developing lung cancer. People who were subjected to Cd at relatively substantial doses in commercial or severely contaminated environments have all suffered from these effects, according to documentation (Bernard, 2008). Due to increased oxidative stress, which is brought on by the formation of reactive oxygen species, biological accumulation of Cd in the body of humans affects the immune system's antioxidant defense mechanism, which in turn leads to many diseases (Suhani et al., 2021). It harms every organ in our body without exception, and the major tragedies it causes are kidney damage, ototoxicity, fatal harm, hormonal difficulties, liver damage, and, above all, cancer risk (Bhattacharyya et al., 2021).

Numerous residential and commercial applications make use of copper (Cu) and its alloys. In the diet of mammals, copper is also a crucial component. The relationship between dose and response is U-shaped, yet its exact shape is still unknown because both copper shortage and copper overload have harmful consequences on health (Stern et al., 2007). A lack of copper can influence the circulatory system, vessels, the pancreas, and other organs' functioning as well as the causes and progression of various digestive and neurologic illnesses. Early in pregnancy, copper deficiency can result in serious organ deformities in the developing fetus. If this condition persists, the child may be born with neurologic and immune problems. However, the existence of significant levels of Cu also poses a serious threat to human health. Acute copper poisoning puts people at risk for a number of pathological disorders and, in extreme situations, can be fatal. Anemia, liver damage, and severe neurological disorders are all results of prolonged copper exposure (Karim et al., 2018).

6.3 Chromium removal by electrodeionization

Chromium is frequently released with the wastewater in the manufacturing sector, which causes major environmental damage. There are several different treatment methods that have been established for eliminating chromium from effluent (Zhao et al., 2018). When an ion exchange and ED techniques are coupled, EDI with anion exchange resin (AER) shows that the ion elimination is not an additive procedure. The outcomes show that the resin's only function is to keep the dilute compartment (DLC) conductive throughout the operation. However, the EDI permits a maximum elimination in less time when a mixed bed exchange resin (MER) is utilized, mostly because of a rise in the number of contact sites between AER and cation exchange resin (CER) material. This approach demonstrated the efficiency of the EDI for eliminating Cr(IV) from an aqueous environment at pH 5 using a MER (Alvarado et al., 2009). Chromium removal was improved when the AER was used in continuous ion exchange together with the Amberlite 200C CER. More than 98.5% of the Cr(VI) was continually eliminated through the EDI procedure at applied current of about 6.6 mA when the MER bed was fully saturated. Thus, it is conceivable that effluent streams comprising Cr(VI) could be employing a combined system, ongoing treatment, EDI. The suggested system may employ an apparatus with a MER bed that is first powered in continues exchange of ions operation and then provided a 10% excess limiting current. This would make it easier to electroregenerate the resin in place and continuously remove chromium from the DLC while recovering chromium for regeneration in the CC (Alvarado et al., 2013). Table 6.1 gives the information about the chromium removal by EDI.

Ionic conduction via the solid component was only possible when AERs were used in the DLC; otherwise, cations and the rest of the anions were transported via the aqueous phase, creating ionic transport rates that are out of balance. If a MER is used, both cations and anions will have access to the solid channel, resulting in a conductance that is far greater than what would be attained in solution and an equal transport of charges velocity. However, MER provides extra locations for in situ renewal procedures, which helps to speed up operation in EDI devices (Alvarado et al., 2015). By testing the I−V relationship, it was effectively discovered that the limiting current was 11.5 mA, 15.0 mA for the unexpanded cell and expanded cell correspondingly. The optimum current was discovered to be approximately 100% and 110% of the maximum current when used with filled resins and IEM. Fresh IERs were first used to examine the impact of unsaturated resins. The elimination of Cr(VI) and Cr(III) by the EDI method was shown to be extremely successful, with removal efficiencies ranging from 95.5% to 98.5% and from 90.0% to 99.0%, correspondingly, within 60 minutes. However, after every run, both resins got more saturated, which decreased the removal effectiveness. The cell voltage, on the other hand, gradually dropped as more cycles were run because the resins got more conductive and assisted in reducing the cell potential. The effectiveness of removal and utilization of energy remained steady when the resins were completely saturated according to additional treatments, although the elimination rate of Cr(VI) was more effective than Cr(III) mostly because of the total charge of Cr(VI) and Cr(III) species (Zhang et al., 2014a).

6.4 Arsenic removal by electrodeionization

Arsenic contamination of water used for drinking in many nations is one of the biggest global public health worries. A

Table 6.1 Removal of chromium by electrodeionization.

Equipment	Component	Operating condition	Removal percentage	References
4-Compartment EDI	2-CEM 1-AEM MER (1:1)	Power consumption: 0.07 kWh/m^3 Initial concentration: 100 ppm Flow rate: 4.5 mL/min Duration: 8 h Voltage: 2.75 V	98.5	Alvarado et al. (2013)
4-Compartment EDI	2-CEM 1-AEM MER (1:1)	Power consumption: 0.167 kWh/m^3 Initial concentration: 100 ppm Flow rate: 12 mL/min Duration: 11 h pH: 7.97	99.8	Alvarado et al. (2009)
5-Compartment EDI	2-CEM 2-AEM MER (1:1)	Initial concentration: 100 ppm Flow rate: 4.5 mL/min Duration: 2 h pH: 3 Voltage: 2.5 V	99.9	Zhang et al. (2014a)
3-Compartment EDI	2-CEM AER	Power consumption: 1.1 kWh/mol Initial concentration: 50 ppm Flow rate: 1.2 L/h Duration: 2 h Voltage: 24.4 V	99	Jina et al. (2020)

(*Continued*)

Table 6.1 (Continued)

Equipment	Component	Operating condition	Removal percentage	References
3-Compartment EDI	1-CEM 1-AEM AER	Power consumption: 4.1 kWh/mol Initial concentration: 50 ppm Flow rate: 0.5 L/h Duration: 40 h Voltage: 22.2 V	99.5	Xing et al. (2009)
3-Compartment EDI	2-AEM AER	Power consumption: 72.39 kWh/kg Initial concentration: 300 ppm Flow rate: 10.7 mL/min Duration: 37.5 h Voltage: 12.5 V	98.82	Basha et al. (2008)
3-Compartment EDI	2-AEM AER	Initial concentration: 50 ppm Flow rate: 5.5 mL/min Duration: 10 h Voltage: 25 V pH: 4.6	99	Dharnaik and Ghosh (2014)

AER, Anion exchange resin; *CER*, cation exchange resin; *MER*, mixed bed exchange resin; *AEM*, anion exchange membrane; *CEM*, cation exchange membrane; *DLC*, dilute compartment; *CC*, concentrate compartment.

5-compartment unit was employed, including two neighboring chambers containing CEM on the ends, a center chamber packed with AER and bordered with AEM, and two rinse chambers, one at each electrode. Arsenic solutions were circulated through the center chamber and the two surrounding chambers as part of the batch-wise EDI operation. Based on pH change, conductance, and As dosage in each section, the arsenic ions in the mixture are swapped by hydroxide in the IER, which is renewed with hydroxide ions created by dissociation of H_2O at the surface of the membrane and electrolyte (Ortega et al., 2017).

Water breakup is essential for IER renewal, a procedure whose strength increases as the cell voltage increases, even at the normally low levels of arsenic in the elimination procedures. Furthermore, the irregular resin renewal caused by the nonhomogeneous current distribution is influenced by design and operating features. The EDI model is used to simulate the laboratory batch behavior of an EDI cell that operates in circulation mode with individual vessels attached to every chamber. The arsenic level varied from 13.3 mg/L to less than 0.01 in DLC, at a constant current density of 8.4 A/m^2 (Rivero et al., 2018).

In batch circulation mode, EDI has the ability to eliminate arsenic from aqueous solutions at strengths varying from 5 to 25 ppm, and in continuous column study, it can do so at 5−15 ppm. Despite operating at a voltage varies between 5 and 20 V, the equipment showed a highest ion percentage elimination rate of nearly 100%. The breakthrough curves were established using a series of column investigations using dosages between 5 and 15 mg/L, voltages between 5 and 20 V, and flow ranges from 5 to 20 mL/min. Arsenic was eradicated up to 98.8%, with energy consumption in the EDI unit varying between 3.88 and 60.7 kWh/kg of eliminated As (Rathi and Kumar, 2022). Lee et al. (2017) demonstrated the behavior of ions during operation in an EDI cell in terms of both movement and exchange. The IER saturation level, cell voltage, and cell age were the three key elements affecting the effectiveness of the EDI operation under the initial feed dosage condition. While the fresh and saturated cells responded to an increase in voltage in an unstable and quick manner, the aged cell displayed consistent and gradual alterations. However, the younger cell stabilized more quickly than the older cell, and it also performed better following stabilization.

6.5 Cobalt removal by electrodeionization

The main radionuclide that dissolves in the main coolant water is the cobalt ion. A two-stage EDI method was evaluated for the elimination of Co ions and for the generation of clean water. It was discovered that over 97% of the cobalt ions could be removed. Despite the IER's variable conductivity, no appreciable variations were found when comparing efficiency of elimination and cell voltage decreases for IER filled in the EDI. The following step of the EDI system, which was filled with a MER, cleansed the effluent from the DLC in the initial stage (Yeon et al., 2003). Continuous EDI (CEDI) procedure used IER, immobilized ion-exchange polyurethane resin (IEPU), and ion-exchange textile (IET) as three ion-carrying spacers. The CER and AER were stacked on top of the CEDI stack to form a bed. The stack arrangement was created to stop the metal ions and hydroxide ions from reacting. At current efficiencies that vary from 18 to 24%, the CEDI operations with the multilayer bed demonstrated above 99% ion elimination (Yeon and Moon, 2003).

CER, AER, and MER were used as a layered bed on which to run the CEDI system. In order to avoid an interaction between HM and OH^- ions, the stack arrangement was created. At 30% of the current efficiency, the excess of 99% of the ions were removed by the EDI method using the multilayer bed. The elimination of HMs at extremely low concentrations via the CEDI operation was therefore effectively proven by this investigation (Yeon et al., 2004b). When synthetic cobalt plating rinse water with 300 ppm cobalt ions was subjected to the electrostatic shielding based coupled EDI treatment, a small quatity of cobalt concentrate that could be returned to the electroplating bath for reuse and a dilute with 43.8 ppm cobalt ions were created. The next electrostatic shielding EDI procedure employed the diluate as feed to generate clean water with a Co ion level below 0.1 ppm. Then, 22%–29% was the current efficiency. The suggested membrane-free electrostatic shielding EDI technique might be developed into a new, cutting-edge electrochemical method for removing cobalt or additional from wastewater from industries as well as for purifying water and reusing it (Dermentzis et al., 2010).

The main coolant at a nuclear power plant can be decontaminated favorably using EDI. In principle, the nuclide may be successfully removed without wasted resin in unit operation, facilitating the reduction of radioactive waste in nuclear waste. In the EDI unit, weak-base AER significantly outperforms the original forms of AER in terms of cobalt adsorption. The cobalt elimination can be enhanced by moving the AER in the CEDI stack. Cobalt concentrations in the modified module's DLC are substantially below than those of the commercial unit, which range from 205 to 289 ng/L and are under the permissible level of ICP-MS measurement of 2 ng/L (Li et al., 2010).

6.6 Nickel removal by electrodeionization

HM nickel is poisonous, and nickel compounds, including nickel sulfide, are thought to be carcinogenic. Up to 1000 ppm of nickel may be present in streams of waste from nickel plating businesses, textile businesses, or cleaning effluents used to clean up nickel-contaminated soil. We looked at producing pure water and recovering nickel from diluted $NiSO_4$ that is the solution using an enhanced EDI technique. It was said that feed water with 50 ppm nickel yielded a nickel rejection of more than 99.8% even though the nickel concentration in the concentrate stream reached 1583 ppm. The concentrate factor might increase to 31.7 and the present efficiency might reach 40% (Lu et al., 2010). Ion movement rises concurrently with a rise in flow rate via the IER, most likely as a consequence of enhanced packing. The IER that had previously been loaded with nickel was examined using the EDI method of removing nickel from a solution that was diluted. It was discovered that nickel ions are carried via both the IER and solution. The degree of solution cleansing was 40% (Dzyazko, 2006).

In membrane-less EDI purposes, electrostatically protected zones–ionic current sinks composed of powdered graphite can act as ion traps and ion CC by eliminating the electric field that is generated inside their weight. The suggested novel EDI cells do not show the existing membrane-related constraints including concentration polarization, scaling, or fouling of the membrane. They may additionally function in extreme chemical or temperature environments. A batch study cleaning of prepared nickel solution including 100 ppm nickel was carried out using a membrane-less electrostatic shielding EDI cell under a fixed voltage and declining current at a current density of $10-20$ A/m^2. Nickel levels dropped during the course of 35 minutes, reaching 17 ppm. Additionally, the continual purification of a 0.001 M nickel sulfate solution into clean water with a nickel content of below 0.1 ppm dilute solution and a current density of 30 A/m^2 was accomplished using a novel membrane-free electrostatic shielding EDI cell. To assist in the completion of environmental or other technical tasks, additional research is required to learn more about the application potential of the suggested ICS and their dynamics (Dermentzis, 2010).

6.7 Other metal ions removal by electrodeionization

The consumption of copper in the electroplating industry is quite common, and its ability to cause cancer in animals has been well established. To avoid the buildup of bivalent metal hydroxide while the EDI cell is functioning, a better arrangement of the membrane's stack was employed. The outcome shows that for a greater elimination effect, a slight drop in the pH value of inlet and a small raise in the supplied voltage were required. With an effectiveness of removal of more than 99.5% and an enrichment factor of 14, the continuous process of electroplating effluent treatment could be attained. Less than 0.23 mg/L of copper was present in filtered water. This showed how HM ions and clean water might be recovered from plating wastewater for use in industrial settings (Feng et al., 2008). Table 6.2 gives the information of various HM removal by EDI. The membrane-free EDI system with a graphene composite electrode demonstrated an exceptional level of copper and pure water recovery and created very little toxic waste, which is important for the ecosystem and the economy. The technology also demonstrated great electric renewal efficiency, reducing energy consumption and wastewater output while avoiding issues with IEM. The application of membrane-free EDI to the purification of effluent containing copper should rise as a result of our results, which demonstrated the ability of the graphite composite electrodes membrane-free EDI system to achieve very effective copper elimination and reclamation from wastewater (Shen et al., 2022).

A four-chamber EDI technique with continuous current was used to treat artificial low radioactive waste water comprising cesium. Investigations are conducted into the impact of different operation factors, including feed rate, electrical current, pH value, starting dosage, and the decline in volume factor. The electrical current flowing through the feed is 0.14 A. Under these circumstances, a 99.9% removal of cesium is

Table 6.2 Removal of other metal ions by electrodeionization.

Metal Ions	Equipment	Component	Operating Condition	Removal Percentage	References
Arsenic	5-Compartment EDI	2-CEM 2-AEM MER	Power consumption: 7.5 kWh/kg Initial Concentration: 15 ppm Flow rate: 30 mL/min Voltage: 15 V pH: 7.3	99.5	Ortega et al. (2017)
Arsenic	5-Compartment EDI	2-CEM 2-AEM AER	Initial concentration: 13.3 mg/L Flow rate: 30 mL/min Voltage: 1.6 V	99.9	Rivero et al. (2018)
Arsenic	3-Compartment EDI	2-AEMAER	Initial concentration: 15 ppm Flow rate: 0.3 L/h Voltage: 20 V	99.9	Rathi et al. (2021)
Arsenic	3-Compartment EDI	2-AEM AER	Power consumption: 3.88 kWh/kg Initial concentration: 15 ppm Flow rate: 20 mL/min Voltage: 20 V	99.9	Rathi et al. (2021)

(Continued)

Table 6.2 (Continued)

Metal Ions	Equipment	Component	Operating Condition	Removal Percentage	References
Arsenic	3-Compartment EDI	1-AEM 1-CEM MER (1:1)	Initial concentration: 20 ppm Flow rate: 10 mL/min Voltage: 5.2 V	99.9	Rathi et al. (2021)
Cobalt	5-Compartment EDI	3-CEM 1-AEM MER	Power Consumption: 1.2CWh/L Initial Concentration: 100 mg/L Flow rate: 5 mL/min Voltage: 20 V pH: 5.3	99	Yeon and Moon (2003)
Cobalt	5-Compartment EDI	3-CEM 1-AEM IET	Power consumption: 9 Wh/L Initial concentration: 100 mg/L Flow rate: 5 mL/min Voltage: 20 V pH: 5.3	99	Yeon and Moon (2003)
Cobalt	5-Compartment EDI	3-CEM 1-AEM IPEU	Power consumption: 2.4 Wh/L Initial concentration: 100 mg/L Flow rate: 5 mL/min Voltage: 20 V pH: 5.3	99	Yeon and Moon (2003)

Cobalt	5-Compartment EDI	2-CEM 2-AEM MER	Power consumption: 14 Wh/L Initial concentration: 0.34-Mm Flow rate: 5 mL/min pH: 5.8	99.9	Yeon et al. (2004b)
Cobalt	Membrane free electrostatic shielding EDI	Ionic current Sinks MER	Initial concentration: 100 ppm Flow rate: 4.06×10^{-4} dm^3/s pH: 3	99	Dermentzis et al. (2010)
Cobalt	7-Compartment EDI	3-CEM 3-AEM MER	Initial concentration: 5000 ng/L Flow rate: 30 L/h Time: 7.27 h Voltage:0−60 V	99	Li et al. (2010)
Strontium	7-Compartment EDI	3-CEM 3-AEM MER	Initial concentration: 5000 ng/L Flow rate: 30 L/h Time: 7.27 h Voltage:0−60 V	98	Li et al. (2010)
Cobalt	5-Compartment EDI	2-CEM 2-AEM MER	Initial concentration: 10 ppm Flow rate: 2 mL/min Time: 8 h	100	Lee et al. (2007)

(Continued)

Table 6.2 (Continued)

Metal Ions	Equipment	Component	Operating Condition	Removal Percentage	References
Cobalt	5-Compartment EDI	2-CEM 2-AEM MER	Initial concentration: 100 mg/L Flow rate: 5 mL/min pH: 5.3	98	Yeon et al. (2004a)
Nickel	13-Compartment EDI	6-CEM 6-AEM MER	Initial concentration: 50 ppm Flow rate: 6 L/h pH: 5.7 Voltage: 12 V Time: 8 h	99.8	Lu et al. (2010)
Nickel	3-Compartment EDI	1-CEM 1-AEM CER	Initial concentration: 300 mol/m^3 Flow rate: 0.21 cm^3/s pH: 2.9 Voltage: 10 V Time: 5 h	99.8	Dzyazko (2006)
Nickel	Membrane free electrostatic shielding EDI	Ionic current Sinks MER	Initial concentration: 100 ppm Flow rate: 2.02×10^{-4} dm^3/s pH: 4 Voltage: 10 V Time: 10 h	99.9	Dermentzis (2010)

Copper	5-compartment EDI	2-CEM 2-AEM MER	Initial concentration: 50 ppm Flow rate: 3 L/h pH: 5.72 Voltage: 30 V Time: 8 h	99.5	Feng et al. (2008)
Copper	Membrane free EDI	Ionic current Sinks MER	Power consumption: 1.48 kWh/m^3 Initial concentration: 50 ppm Flow rate: 7.1 L/h pH: 7.37 Voltage: 1.6 V Time: 2.5 h	99	Shen et al. (2022)
Copper	Membrane free EDI	Ionic current Sinks MER	Initial concentration: 100 ppm Flow rate: 1.29×10^{-4} L/s pH: 2.5 Voltage: 50 V Time: 35 min	84.2	Dermentzis et al. (2009)
Cesium	4-Compartment EDI	2-CEM 2-AEM MER	Initial concentration: 50 ppm Flow rate: 6 L/h pH: 7 Time: 8 h	99.9	Zhang et al. (2014b)

AER, Anion exchange resin; *CER*, cation exchange resin; *MER*, mixed bed exchange resin; *AEM*, anion exchange membrane; *CEM*, cation exchange membrane; *DLC*, dilute compartment; *CC*, concentrate compartment.

possible and the waste water cesium level in the DLC is under 0.05 ppm (Zhang et al., 2014b). Dermentzis et al. (2011) studied that electrode graphite powder-based electrostatic shielding regions were built and employed as a novel kind of ICS. Ions circulate within these regions and aggregate there as a result of the localized removal of the electrical field that is applied, voltage, and current inside them. EDI of synthetic Cd effluent containing 50 ppm cadmium ions while simultaneously electrochemically renewing the IER beds with the ICS. At a flow rate of 3.27×10^4 L/s DLC, a current density of 2 mA/cm^2, and a current efficiency of 28% for cadmium ion elimination, clean water was achieved with a Cd ion level of below 0.1 ppm.

6.8 Conclusion

EDI has recently been shown to be a successful technique for eliminating ionic compounds from contaminated rivers. This chapter provides a concise summary of HMs, their causes, and their effects. It has also assessed several methods of EDI-based HM removal. With the help of the offered applications examples, it has been shown that the EDI technique is quite efficient in recovering some important species, such HMs. Due to EDI, a very high percentage of these pollutants are consistently eliminated. As of now, EDI has excelled as a very effective, ecologically friendly process of cleansing, separation, and concentration. Primary benefits of technique include sludge-free procedure, economical, and exceptional ionic removal efficiency. With other procedures, it is not possible to extract HM from wastewater at even low concentrations. EDI without a membrane and electrostatic shielding using novel substances for HM removal were also addressed. It is projected that EDI systems would continue to advance, which will make them more appealing for industrial scale-up across a variety of uses throughout the world since they operate at cheaper prices due to reduced energy demand.

References

Alvarado, L., Ramírez, A., Rodríguez-Torres, I., 2009. Cr (VI) removal by continuous electrodeionization: study of its basic technologies. Desalination 249 (1), 423–428.

Alvarado, L., Torres, I.R., Chen, A., 2013. Integration of ion exchange and electrodeionization as a new approach for the continuous treatment of hexavalent chromium wastewater. Separation and Purification Technology 105, 55–62.

Alvarado, L., Rodríguez-Torres, I., Balderas, P., 2015. Investigation of current routes in electrodeionization system resin beds during chromium removal. Electrochimica Acta 182, 763—768.

Arar, Ö., Yüksel, Ü., Kabay, N., Yüksel, M., 2014. Various applications of electrodeionization (EDI) method for water treatment—a short review. Desalination 342, 16—22.

Asati, A., Pichhode, M., Nikhil, K., 2016. Effect of heavy metals on plants: an overview. International Journal of Application or Innovation in Engineering & Management 5 (3), 56—66.

Assi, M.A., Hezmee, M.N.M., Sabri, M.Y.M., Rajion, M.A., 2016. The detrimental effects of lead on human and animal health. Veterinary world 9 (6), 660.

Baruthio, F., 1992. Toxic effects of chromium and its compounds. Biological Trace Element Research 32, 145—153.

Basha, C.A., Ramanathan, K., Rajkumar, R., Mahalakshmi, M., Kumar, P.S., 2008. Management of chromium plating rinsewater using electrochemical ion exchange. Industrial & Engineering Chemistry Research 47 (7), 2279—2286.

Bernard, A., 2008. Cadmium & its adverse effects on human health. Indian Journal of Medical Research 128 (4), 557—564.

Bhattacharyya, K., Sen, D., Laskar, P., Saha, T., Kundu, G., Ghosh Chaudhuri, A., et al., 2021. Pathophysiological effects of cadmium (II) on human health—a critical review. Journal of Basic and Clinical Physiology and Pharmacology (0), .

Bjørklund, G., Aaseth, J., Chirumbolo, S., Urbina, M.A., Uddin, R., 2018. Effects of arsenic toxicity beyond epigenetic modifications. Environmental Geochemistry and Health 40 (3), 955—965.

Bridges, C.C., Zalups, R.K., 2017. The aging kidney and the nephrotoxic effects of mercury. Journal of Toxicology and Environmental Health, Part B 20 (2), 55—80.

Briffa, J., Sinagra, E., Blundell, R., 2020. Heavy metal pollution in the environment and their toxicological effects on humans. Heliyon 6 (9), e04691.

Ceja-Galicia, Z.A., Daniel, A., Salazar, A.M., Pánico, P., Ostrosky-Wegman, P., Díaz-Villaseñor, A., 2017. Effects of arsenic on adipocyte metabolism: is arsenic an obesogen? Molecular and Cellular Endocrinology 452, 25—32.

Debnath, B., Singh, W.S., Manna, K., 2019. Sources and toxicological effects of lead on human health. Indian Journal of Medical Specialities 10 (2), 66.

Demayo, A., Taylor, M.C., Taylor, K.W., Hodson, P.V., Hammond, P.B., 1982. Toxic effects of lead and lead compounds on human health, aquatic life, wildlife plants, and livestock. Critical reviews in environmental science and technology 12 (4), 257—305.

Dermentzis, K., Davidis, A., Papadopoulou, D., Christoforidis, A., Ouzounis, K., 2009. Copper removal from industrial wastewaters by means of electrostatic shielding driven electrodeionization. Journal of Engineering Science & Technology Review 2 (1).

Dermentzis, K., 2010. Removal of nickel from electroplating rinse waters using electrostatic shielding electrodialysis/electrodeionization. Journal of Hazardous Materials 173 (1—3), 647—652.

Dermentzis, K.I., Davidis, A.E., Dermentzi, A.S., Chatzichristou, C.D., 2010. An electrostatic shielding-based coupled electrodialysis/electrodeionization process for removal of cobalt ions from aqueous solutions. Water Science and Technology 62 (8), 1947—1953.

Dermentzis, K., Christoforidis, A., Papadopoulou, D., Davidis, A., 2011. Ion and ionic current sinks for electrodeionization of simulated cadmium plating rinse waters. Environmental Progress & Sustainable Energy 30 (1), 37—43.

Dharnaik, A.S., Ghosh, P.K., 2014. Hexavalent chromium [Cr (VI)] removal by the electrochemical ion-exchange process. Environmental Technology 35 (18), 2272—2279.

Duruibe, J.O., Ogwuegbu, M.O.C., Egwurugwu, J.N., 2007. Heavy metal pollution and human biotoxic effects. International Journal of physical sciences 2 (5), 112—118.

Dzyazko, Y.S., 2006. Purification of a diluted solution containing nickel using electro-deionization. Desalination 198 (1–3), 47–55.

Emenike, E.C., Iwuozor, K.O., Anidiobi, S.U., 2021. Heavy metal pollution in aquaculture: sources, impacts and mitigation techniques. Biological Trace Element Research 1–17.

Feng, X., Gao, J.S., Wu, Z.C., 2008. Removal of copper ions from electroplating rinse water using electrodeionization. Journal of Zhejiang University-SCIENCE A 9 (9), 1283–1287.

Fu, F., Wang, Q., 2011. Removal of heavy metal ions from wastewaters: a review. Journal of Environmental Management 92 (3), 407–418.

Genchi, G., Carocci, A., Lauria, G., Sinicropi, M.S., Catalano, A., 2020. Nickel: human health and environmental toxicology. International Journal of Environmental Research and Public Health 17 (3), 679.

Ha, E., Basu, N., Bose-O'Reilly, S., Dórea, J.G., McSorley, E., Sakamoto, M., et al., 2017. Current progress on understanding the impact of mercury on human health. Environmental Research 152, 419–433.

He, B., Yun, Z., Shi, J., Jiang, G., 2013. Research progress of heavy metal pollution in China: sources, analytical methods, status, and toxicity. Chinese Science Bulletin 58, 134–140.

Honda, S.I., Hylander, L., Sakamoto, M., 2006. Recent advances in evaluation of health effects on mercury with special reference to methylmercury—a minireview. Environmental Health and Preventive Medicine 11 (4), 171–176.

Hu, H., Jin, Q., Kavan, P., 2014. A study of heavy metal pollution in China: current status, pollution-control policies and countermeasures. Sustainability 6 (9), 5820–5838.

Jacob, J.M., Karthik, C., Saratale, R.G., Kumar, S.S., Prabakar, D., Kadirvelu, K., et al., 2018. Biological approaches to tackle heavy metal pollution: a survey of literature. Journal of Environmental Management 217, 56–70.

Järup, L., 2003. Hazards of heavy metal contamination. British Medical Bulletin 68 (1), 167–182.

Jensen, A.A., Tuchsen, F., 1990. Cobalt exposure and cancer risk. Critical Reviews in Toxicology 20 (6), 427–439.

Jina, Q., Yaob, W., Chenc, X., 2020. Removal of Cr (VI) from wastewater by simplified electrodeionization. Desalination and Water Treatment 183, 301–306.

Karim, N., 2018. Copper and human health—a review. Journal of Bahria University Medical and Dental College 8 (2), 117–122.

Kumar, P.S., Varsha, M., Rathi, B.S., Rangasamy, G., 2022. Electrodeionization: Fundamentals, methods and applications. Environmental Research 114756.

Lee, J.W., Yeon, K.H., Song, J.H., Moon, S.H., 2007. Characterization of electroregeneration and determination of optimal current density in continuous electrodeionization. Desalination 207 (1–3), 276–285.

Lee, D., Lee, J.Y., Kim, Y., Moon, S.H., 2017. Investigation of the performance determinants in the treatment of arsenic-contaminated water by continuous electrodeionization. Separation and Purification Technology 179, 381–392.

Leyssens, L., Vinck, B., Van Der Straeten, C., Wuyts, F., Maes, L., 2017. Cobalt toxicity in humans—a review of the potential sources and systemic health effects. Toxicology 387, 43–56.

Li, F.Z., Zhang, M., Zhao, X., Hou, T., Liu, L.J., 2010. Removal of Co2 + and Sr2 + from a primary coolant by continuous electrodeionization packed with weak base anion exchange resin. Nuclear Technology 172 (1), 71–76.

Li, C., Zhou, K., Qin, W., Tian, C., Qi, M., Yan, X., et al., 2019. A review on heavy metals contamination in soil: effects, sources, and remediation techniques. Soil and Sediment Contamination: An International Journal 28 (4), 380–394.

Lu, H., Wang, J., Yan, B., Bu, S., 2010. Recovery of nickel ions from simulated electroplating rinse water by electrodeionization process. Water Science and Technology 61 (3), 729–735.

Malik, L.A., Bashir, A., Qureashi, A., Pandith, A.H., 2019. Detection and removal of heavy metal ions: a review. Environmental Chemistry Letters 17, 1495–1521.

Mistry, G., Popat, K., Patel, J., Panchal, K., Ngo, H.H., Bilal, M., et al., 2022. New outlook on hazardous pollutants in the wastewater environment: occurrence, risk assessment and elimination by electrodeionization technologies. Environmental Research 115112.

Ortega, A., Oliva, I., Contreras, K.E., González, I., Cruz-Díaz, M.R., Rivero, E.P., 2017. Arsenic removal from water by hybrid electro-regenerated anion exchange resin/electrodialysis process. Separation and Purification Technology 184, 319–326.

Paul, D., 2017. Research on heavy metal pollution of river Ganga: a review. Annals of Agrarian Science 15 (2), 278–286.

Paul, D.S., Hernández-Zavala, A., Walton, F.S., Adair, B.M., Dědina, J., Matoušek, T., et al., 2007. Examination of the effects of arsenic on glucose homeostasis in cell culture and animal studies: development of a mouse model for arsenic-induced diabetes. Toxicology and Applied Pharmacology 222 (3), 305–314.

Rathi, B.S., Kumar, P.S., Ponprasath, R., Rohan, K., Jahnavi, N., 2021. An effective separation of toxic arsenic from aquatic environment using electrochemical ion exchange process. Journal of Hazardous Materials 412, 125240.

Rathi, B.S., Kumar, P.S., 2022. Continuous electrodeionization on the removal of toxic pollutant from aqueous solution. Chemosphere 291, 132808.

Rathi, B.S., Kumar, P.S., Parthiban, R., 2022. A review on recent advances in electrodeionization for various environmental applications. Chemosphere 289, 133223.

Rivero, E.P., Ortega, A., Cruz-Díaz, M.R., González, I., 2018. Modelling the transport of ions and electrochemical regeneration of the resin in a hybrid ion exchange/electrodialysis process for As (V) removal. Journal of Applied Electrochemistry 48, 597–610.

Rodriguez, V.M., Jiménez-Capdeville, M.E., Giordano, M., 2003. The effects of arsenic exposure on the nervous system. Toxicology Letters 145 (1), 1–18.

Shen, X., Liu, Q., Li, H., Kuang, X., 2022. Membrane-free electrodeionization using graphene composite electrode to purify copper-containing wastewater. Water Science & Technology 86 (7), 1733–1744.

Shrivastava, R., Upreti, R.K., Seth, P.K., Chaturvedi, U.C., 2002. Effects of chromium on the immune system. FEMS Immunology and Medical Microbiology 34 (1), 1–7.

Sodango, T.H., Li, X., Sha, J., Bao, Z., 2018. Review of the spatial distribution, source and extent of heavy metal pollution of soil in China: impacts and mitigation approaches. Journal of Health and Pollution 8 (17), 53–70.

Stern, B.R., Solioz, M., Krewski, D., Aggett, P., Aw, T.C., Baker, S., et al., 2007. Copper and human health: biochemistry, genetics, and strategies for modeling dose-response relationships. Journal of Toxicology and Environmental Health, Part B 10 (3), 157–222.

Suhani, I., Sahab, S., Srivastava, V., Singh, R.P., 2021. Impact of cadmium pollution on food safety and human health. Current Opinion in Toxicology 27, 1–7.

Sundseth, K., Pacyna, J.M., Pacyna, E.G., Pirrone, N., Thorne, R.J., 2017. Global sources and pathways of mercury in the context of human health. International Journal of Environmental Research and Public Health 14 (1), 105.

Tchounwou, P.B., Yedjou, C.G., Patlolla, A.K., Sutton, D.J., 2012. Heavy metal toxicity and the environment. Molecular, Clinical and Environmental Toxicology 3, 133–164.

Tyler, C.R., Allan, A.M., 2014. The effects of arsenic exposure on neurological and cognitive dysfunction in human and rodent studies: a review. Current Environmental Health Reports 1, 132–147.

Xing, Y., Chen, X., Yao, P., Wang, D., 2009. Continuous electrodeionization for removal and recovery of Cr (VI) from wastewater. Separation and Purification Technology 67 (2), 123−126.

Yeon, K.H., Moon, S.H., 2003. A study on removal of cobalt from a primary coolant by continuous electrodeionization with various conducting spacers. Separation Science and Technology 38 (10), 2347−2371.

Yeon, K.H., Seong, J.H., Rengaraj, S., Moon, S.H., 2003. Electrochemical characterization of ion-exchange resin beds and removal of cobalt by electrodeionization for high purity water production. Separation Science and Technology 38 (2), 443−462.

Yeon, K.H., Song, J.H., Moon, S.H., 2004a. Preparation and characterization of immobilized ion exchange polyurethanes (IEPU) and their applications for continuous electrodeionization (CEDI). Korean Journal of Chemical Engineering 21, 867−873.

Yeon, K.H., Song, J.H., Moon, S.H., 2004b. A study on stack configuration of continuous electrodeionization for removal of heavy metal ions from the primary coolant of a nuclear power plant. Water Research 38 (7), 1911−1921.

Zambelli, B., Ciurli, S., 2013. Nickel and human health. Interrelations between Essential Metal Ions and Human Diseases. pp. 321−357.

Zambelli, B., Uversky, V.N., Ciurli, S., 2016. Nickel impact on human health: an intrinsic disorder perspective. Biochimica et Biophysica Acta (BBA)-Proteins and Proteomics 1864 (12), 1714−1731.

Zhang, Q., Wang, C., 2020. Natural and human factors affect the distribution of soil heavy metal pollution: a review. Water, Air, & Soil Pollution 231, 1−13.

Zhang, Z., Liba, D., Alvarado, L., Chen, A., 2014a. Separation and recovery of Cr (III) and Cr (VI) using electrodeionization as an efficient approach. Separation and Purification Technology 137, 86−93.

Zhang, Y., Wang, L., Xuan, S., Lin, X., Luo, X., 2014b. Variable effects on electrodeionization for removal of Cs + ions from simulated wastewater. Desalination 344, 212−218.

Zhao, Y., Kang, D., Chen, Z., Zhan, J., Wu, X., 2018. Removal of chromium using electrochemical approaches: a review. International Journal of Electrochemical Science 13, 1250−1259.

CHAPTER SEVEN

Electrodeionization in desalination and water softening

7.1 Introduction

Despite being a plentiful resource of nature, 97% of the water that exists on earth is saltwater and is unable to be directly ingested. The increasing modernization and rise in population growth have further raised the need for clean water. Water shortage is now one of the major issues facing many nations, and it is predicted that this problem will only get worse in the coming years. Desalination (DS) methods are frequently employed to supply the need for clean drinking water, particularly in locations where drinking water resources are a handful and seawater or saltwater is readily accessible (Sevda et al., 2015). DS, which includes the elimination of pollutants, sodium chloride, and other unwanted ions, is widely recognized as a viable alternative method for producing clean drinking water in order to solve the water shortage (Shahmirzadi et al., 2018). Brackish water, saltwater, and brine are aqueous saltwater solutions that can be converted into clean water via the DS process. There are growing worries about potential adverse ecological consequences, despite the fact that the DS process has numerous benefits. Both the establishment and functioning of DS facilities can have an influence on the surrounding environment. DS produces a coproduct termed "brine" that is released into the oceans and has a high level of salinity as well as residues of chemicals. Other significant difficulties, besides brine, include the considerable energy expenditure of DS and brine purification methods, as well as the release of greenhouse gases and airborne contaminants. In addition, there are problems in using chemicals and entrainment and trapping of aquatic creatures (Panagopoulos and Haralambous 2020).

According to Subramani & Jacangelo (2015), DS techniques include membrane-based, electrochemical-based, thermal-based, and other technologies. One of the most essential and well-known methods for DS as well as wastewater treatment is membrane technology that utilizes polymer membranes. Although there are a few identified obstacles with polymer membranes, technological advancement and the related expertise in the fabrication of inorganic membranes continue unavoidably to address some of the root causes (Goh and Ismail 2018). A beneficial method for desalinating waters with salinities in the range of saline water is electrodeionization (EDI). When used together, both electrodialysis (ED) and ion exchange (IE) can overcome their own limits. As water is desalinated, ED exhibits larger electrical resistance in the dilute compartment (DLC). As a result, more energy is needed to produce high-purity water (HPW), which significantly reduces productivity. On the other hand, the IE process cannot run constantly since it has to chemically regenerate depleted ion-exchange resin (IER) in order to recover it. However, the EDI process enables continuous operation under the impact of an electric field that replenishes IER due to the collaboration of two techniques. HPW can therefore be created. Without the requirement for chemical IER regeneration, the electric voltage placed across the electrodes constantly desalinates water. Due to this significant benefit, EDI is an environmentally beneficial method (Otero et al., 2021).

Typical water-related factors such as total alkaline content and overall hardness are crucial for farming output, aquatic system productivity as a whole, and water availability and usage. Alkalinity and hardness data, along with information on a water's total dissolved solid content, may be utilized to arrive at crucial conclusions about how helpful a water is for a variety of uses. To determine whether water will be corrosive or establish an accumulation of deposits that clogs pipelines in water delivery systems, knowledge on the alkalinity and hardness of water sources is necessary. It can be employed as well to calculate the quantities of reagents required for softening the water in basic models to identify the treatments required to prepare the water for usage (Boyd et al., 2016). Several treatments, including microbial and electrochemical processes, electrolysis, adsorption, nanofiltration, IE, and chemical precipitation (CP), can be used to soften water (Seo et al., 2010). It is often possible to attain minimal hardness using a mix of cationic IE cells and lime softening devices. Other techniques using IERs have grown more interesting as continuous techniques with renewal capabilities (Campanile et al., 2022). One of the advantages of EDI reversal (EDIR), which is EDI employing recurrent polarity reversal (PR), is that it has the ability to avoid or significantly reduce fouling of membranes and scaling. The effectiveness of the EDIR method was examined in order to eliminate materials that include divalent cations, which are substances that cause hardness and to demonstrate its viability as a water softening (WS) method (Lee, et al., 2012). The purpose of this chapter was to look at the many approaches of DS and WS, as well as their uses. In addition, recent advancements in the application of EDI in DS and WS have been researched. In terms of cost and environmental consideration, EDI performs better than other approaches.

7.2 Desalination

DS of saltwater and brackish water levels is receiving greater attention since water supplies are quickly running out. DS techniques used today demand a lot of energy, which is expensive in terms of both economics and damage to the environment (Karagiannis and Soldatos 2008). Over the past century, multiple innovations have been suggested. Reverse osmosis (RO), multistage flash (MSF) DS, and multieffect distillation (MED) are the examples of present commercial methods. In these situations, seawater treatment facilities are identical to conventional ones; the only distinction is that renewable energy is used in them. As a result, categories are first established while taking into account the underlying concepts, the primary energy input needed for the procedure, and the possibilities for integrating sources of clean energy (Curto et al., 2021). Fig. 7.1 depicts the different techniques for DS.

DS costs have significantly decreased as a consequence of substantial technical advancements, enabling the water that is desalinated affordable with other sources of water, which in turn has led to an expansion in DS capacity. This growth in capacity is mostly due to both increasing demand for water and the substantial decrease in DS costs. DS is currently able to effectively compete with traditional sources of water and transportation of water for the delivery of drinking water in select specified places (Ghaffour et al., 2013). Environmental concerns associated with DS play a significant role in the development and use of DS technology. An appropriate DS plant is anticipated to comply with ecological standards and be economical to build, operate, and maintain, as well as to pay monitoring and licensing costs. The placement of DS plants and water intake infrastructure, as well as concentrate management and disposal, are among the main environmental considerations (Younos, 2005).

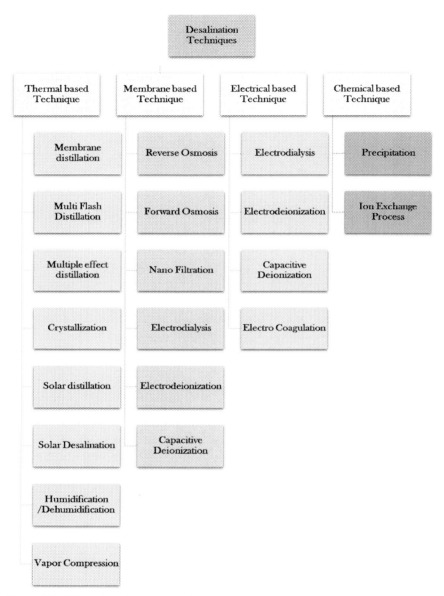

Figure 7.1 Different techniques for desalination.

7.2.1 Different techniques for desalination

Systems for desalinating water can be divided into groups based on the forms of energy they use, including thermal energy, mechanical energy, electrical energy, and chemical sources of energy. Evaporation—condensation, filtering, and crystalizing method make up another categorization that is dependent on the DS procedure. Adsorption DS, membrane DS (MD), ED (EDI), membrane-based bioreactors (MBRs), forward osmosis (FO), and IER are a few DS methods still under research. The most widely used techniques for DS globally are RO, MSF, and MED systems (Alkaisi et al., 2017). Membrane technologies (MT), distillation procedures (thermal techniques), and chemical methods are the three fundamental kinds of water purification technologies utilized for DS. A few water treatment facilities combine these methods. The most prevalent DS technique is MT, although thermal methods are less popular. Chemical methods include procedures such as IE, which are thought to be inefficient for purifying waters with large concentrations of dissolved solids (Younos and Tulou 2005). Table 7.1 represents the efficiency of DS with different techniques along with its operating condition. A list of recent technologies being researched and developed for potential use with DS is also included in this chapter.

7.2.1.1 Solar desalination

Fossil energy prices may be expensive whereas solar power is widely available in many locations where freshwater supplies or water delivery infrastructure are lacking. Solar desalination (SDS) systems are being considered as a complement to current freshwater sources in water-scarce locations due to the industrialized world's government policies, which emphasize replacing fossil fuels with renewable, low-carbon energy sources. SDS is appealing as a way to save energy from fossil fuels and lessen the carbon imprint of DS, even in areas where petroleum supplies are abundant. Finally, in rural and "off-the-grid" locations, an SDS system could be cheaper than options such as trucked-in water or DS powered by diesel power (Antar et al., 2012). There are two SDS system changes discussed. To improve the still's effectiveness, the initial alteration added a packed layer to the basin's bottom. Glass balls are used to create a packed layer that is thought of as a straightforward thermal storage mechanism. The second alteration uses a spinning shaft that is put not far below the surface of the water in the basin. The performance of the current SDS system might be improved with the goal of employing the spinning shaft to breach the outermost layer of the basin's surface water and boost water vaporization and condensation (Abdel-Rehim and Lasheen 2005).

SDS can be direct, using solar energy to make distillate right in the solar collector, or indirect, using MSF distillation, vapor compression (VC), RO,

Table 7.1 Efficiency of desalination (DS) with different techniques along with its operating condition.

Salt removal	Technique	Operating condition	Efficiency for DS	References
NaCl	Capacitive deionization	Initial conductivity: 110 μS/cm Voltage: 1.2 V	60%.	Li and Zou (2011)
NaCl	Capacitive deionization	Initial concentration: 20 mM Current density: 2.5 A/m^2 Flow rate: 6.4–11.8 mL/min Energy consumption: 22 Wh/m^3	70 mg NaCl/ electrodes	Singh et al. (2020)
Na$_2$SO$_4$, NaCl	CDI and photocatalysis	Initial concentration: 10^{-3} mol/L Flow rate: 1 mL/cm^2 min Current density: of 5 mA/cm$_2$	80%	Ye et al. (2019)
NaCl	EDI	Initial concentration: 3 g/L Flow rate: 810 mL/min Energy consumption: 0.66 kWh/m^3 Voltage: 2.28 V	94%	Zheng et al. (2018)
NaCl	EDI	Initial concentration: 0.01 M Voltage: 14 V Energy consumption: 4.54 kWh/m^3 Time: 60 min	73.34%	Otero et al. (2021)
NaCl	EDI	Initial concentration: 0.02 M Voltage: 14 V Energy consumption: 5.38 kWh/m^3 Time: 60 min	63.89%	Otero et al. (2021)
NaCl	Electrodialysis	Initial concentration: 5000 ppm pH: 8.4 Voltage: 12 V Flow rate: 2.5 L/min Time: 60 min	90%	Banasiak et al. (2007)
NaCl	Membrane capacitive deionization	Initial conductivity: 110 μS/cm Voltage: 1.2 V	97%	Li and Zou (2011)
NaCl	Membrane distillation with ceramic planar membrane	Initial concentration: 4–12 wt.% Flow rate: 100 L/h Temperature: 75°C Water flux: 11.75–9.19 L/(m^2 h)	83%	Tao et al. (2018)

NaCl	Membrane-free electrodeionization	Initial conductivity: 50 μS/cm Energy consumption: 1.5 kWh/m^3 Flow rate: 15 m/h Voltage: 1400 V Time: 60 min	86%	Shen et al. (2014)
NaCl	Resin wafer EDI	Initial concentration: 5 g/L Time: 120 min Flow rate: 1100 mL/min	90%	Pan et al. (2017)
NaCl	Resin wafer EDI	Initial concentration: 5 g/L Time: 120 min Flow rate: 1100 mL/min Energy consumption: 0.66 kWh/m^3	99%	Pan et al. (2018)
NaCl	RO	Initial concentration: 25 g/L Time: 10 min	97.2%	Laycock et al. (2012)
NaCl	RO system	Initial concentration: 20,000−35,000 ppm Pressure: 700−1000 psi Temperature: 20°C−30°C Flow rate: 300 cm^3/s	30%	Hawlader et al. (2000)
NaCl	RO/EDI	Initial conductivity: 400 μS/cm Flow rate: 20 L/h	95%	Wang et al. (2000)

CDI, Capacitive deionization; *EDI*, electrodeionization; *RO*, reverse osmosis.

MD, and ED, together with solar collectors for heat production. In comparison with indirect methods, direct SDS takes a lot of land and is less productive. On the other hand, because to its comparative affordable price and the ease of use, it is attractive to the indirect DS plants in small-scale production (Qiblawey and Banat 2008). Although photovoltaic-driven RO seems to be the most advanced technology, the price of water is now far greater than that generated by systems that use fossil fuels. Large-scale plants having access to low-salinity feed water and high amounts of insolation often have the lowest costs (Pugsley et al., 2016).

SDS stands out among other distillation methods as a very appealing and straightforward technology. It is best suited for small-scale operations in areas with abundant sun energy. India enjoys a lot of sunshine because it is a tropical nation. For various regions of the nation, the average daily sun radiation ranges between 4 and 7 kWh/m^2. It receives roughly 5000 trillion kWh of solar energy year due to the 250—300 bright sunny days that occur on average throughout the year. Solar energy has the capacity to satisfy and complement a variety of energy needs, despite the drawbacks of being a diluted source and irregular in nature. Due to its modular design, solar energy systems may be placed in any configuration (Arjunan et al., 2009). In the first, solar energy (SE) is absorbed and employed in a single piece of machinery, whereas in the second, two independent subsystems—one for SE conversion and one for DS—are utilized. Different systems are examined in terms of their main energy consumption, need for treating seawater, cost, and appropriateness for SE utilization. The multiple-effect boiling (MEB) system and, in particular, the multiple-effect stack (MES) type evaporator are among the several types of processes that are investigated, and they are the most appropriate for using SE because they can operate without disruption under a variety of energy sources. This method also needs the most basic seawater treatment and has a low specific consumption of energy and affordable costs for the equipment (Kalogirou 1997).

7.2.1.2 Membrane distillation

MD is a potential thermally powered DS technique that is still in the research and development stage as well as being commercially untapped. A hydrophobic membrane, that is permeable to water vapor but hinders liquid water, is used in the process to purify water. The greater water vapor pressure caused by higher temperatures pushes vapor via the small openings of the hydrophobic membrane in saline DS applications of MD, where it accumulates on the permeate side. Compared to other DS technologies, including as pressure-driven techniques such as RO and thermally-driven techniques such as MSF, MD has distinct benefits. The specialized demands of high-pressure RO systems, including as large gage pipes, complicated pumps, and maintenance demands, are not present in MD. MD offers the possibility of 100% rejections of salts and particles and is more fouling tolerant than RO since it is not a pressure-driven process and just vapor is permitted to pass via the membrane. Because MD has a lot fewer components and operates at lower temperatures compared to other thermal systems, it may possess a smaller overall footprint. This is possible due to the decreased vapor space (Warsinger et al., 2015).

Aside from being small and using low-temperature heat effectively, the MD method also appears to be more immune to clogging than other MT. The following impacts of common MD setups' operational parameters are seen within the studied range: (1) By raising the hot feed temperature and decreasing the vapor/air gap, permeate fluxes may be significantly enhanced. (2) The mass flow velocity of the feed mixture has a lesser effect. (3) The amount of solute has a minor impact. (4) The cold side conditions possess a smaller impact on fluxes than the one on the hot side. (5) The coolant mass flow rate has little impact. (6) The coolant temperature has a smaller impact than the mass flow velocity of the hot solution's temperature (Alklaibi and Lior 2005). Due to their inherent hydrophobic qualities and high accessibility, polymeric membranes have often dominated MD research. On the contrary hand, ceramic membranes for MD DS are being developed and are increasingly replacing their polymeric equivalents because of their higher chemical, mechanical, and thermal stabilities and potential for greater service lives. It is important to note that ceramic membranes made from 1-D nanostructures such as nanowires and nanofibers have drawn attention and show certain superior qualities to deliver exceptional MD performances. More work is required to encourage ceramic membranes as a superior substitute for polymeric membranes in MD DS applications because their progress is still relatively modest in comparison with that of polymeric membranes (Tai et al., 2020).

7.2.1.3 Reverse osmosis

In the last 40 years, RO technique has advanced to the point that it now commands a share of the market of 44% of the DS manufacturing capacity and 80% of the DS plants deployed globally. Over the last 10 years, advancement in the use of traditional and low-pressure membrane preprocessing procedures for saltwater RO DS has increased. Since membrane fouling brought on by particulate matter/colloids, organic/inorganic chemicals, and biological development is a major problem, reliable preprocessing procedures are necessary for the efficient functioning of RO systems. While traditional RO pretreatment procedures such as coagulation and granular medium filtration have been extensively employed, there has been a growing trend toward using ultrafiltration/microfiltration in place of traditional treatment methods (Valavala et al., 2011).

Today, RO, which is powered by a pressure gradient across a semipermeable membrane, is a particularly promising technique for desalinating saltwater. This method has received a lot of interest in part because it uses relatively little energy while operating and is simpler to use and maintain than other established systems such as thermal DS (Misdan et al., 2012). Through repeated simulation, the impact of various operational factors such as feed flow rate, feed pressure, and design characteristics such as internal diameter, total number of tubes, on the recovery ratio is investigated. Finally, a framework for optimization of processes is created to use various devices for energy recovery and maximize the recovery ratio or a profit operation, subject to general restrictions. The best values for the decision variables are determined by the limitations put in place, and they are also sensitive to the changes in feed concentration, water and energy pricing, and both. The usage of newly developed energy recovery devices is often justified since they show substantially greater operational cost reductions than the conventional technology currently in use. Up to 50% less energy can be used when using a pressure exchanger gadget (Villafafila and Mujtaba 2003).

However, more permeability membranes with strong resistance to disinfecting agents and a small fouling tendency should enable reduced plant size and extended membrane lifespan. Additionally, enhancements in RO versatility, energy-optimal plant control and self-adapting procedure, actual time tracking of the membrane's integrity and fouling, lesser plant footprints, improved membrane element design, more effective pumps, and devices for energy recovery can all help to reduce costs. It is emphasized that all of the aforementioned factors are intimately related, and that progress in any one of them is likely to have an effect on the others. While there have been many suggested substitutes to RO DS, nothing have, to far, been demonstrated in field experiments to be more effective than RO DS for huge-scale applications of producing potable water. Future advancements may spur competition, which could ultimately result in the spread and greater acceptance of DS as a crucial step toward water sustainability (Cohen et al., 2017).

In the RO method, the concentrated brine always gets dumped, which is bad for the economy and might also potentially harm the surroundings and ecosystem. In order to reduce brine outflow, improve environmental friendliness, and reap economic rewards, it is effective to increase recovery rates. Inorganic scaling, on the other hand, is brought on by the overabundance of mineral ions as a result of the continually rising concentration and results in a decreasing recovery rate. Scaling made of inorganic materials might obstruct membranes, drastically reduce permeations, or even stop engineering operations. Therefore, improving RO DS's antiscaling capacity is crucial (Liu et al., 2019). Commercial RO membranes were coated with polyethylene glycol-based hydrogels that had been created. The hydrogels provide intriguing candidates for covering materials that are fouling-resistant because of their high-water absorption and permeability to water. Membranes with coatings have a lower water flux than those without coatings (Sagle et al., 2009).

7.2.1.4 Electrodialysis

A very experienced MT, ED, DS, and solution regeneration using ED are employed in a variety of sectors because to its versatility, adaptation to the surroundings, and saving energy capabilities. The mass transfer model is crucial for capturing the internal workings of ED and for optimizing its structure and performance (Liu and Cheng 2020). ED takes up a relatively small amount of seawater DS in comparison with RO. The cost of ED is said to be the only factor allowing ED to compete with RO in the range of water input salinity up to $8-10$ g/L. The amount of salt that must be transferred across the membrane and DS are closely connected (Turek 2003). One of the earliest large-scale water DS technologies based on MT that was commercially accessible was ED, which is still in use today. However, despite its long history of dependable operation and evident technological benefits, RO and NF have been supplanting ED in saltwater DS and wastewater treatment in recent years. The total process costs are the key factor causing a decrease in ED usage in the manufacture of potable water (Strathmann et al., 2004). The invention of ion-exchange membranes (IEMs), which offer high-water recovery without requiring phase shift, reaction, or chemicals, is what drives the functioning of ED. Without using energy from fossil fuels or chemical chemicals, such advantages offer environmental advantages. Scaling, fouling of membranes, and perm selectivity are still issues with ED technology (Sadrzadeh and Mohammadi 2008).

Without the addition of chemicals, ED DS delivers excellent water recovery. Due to its capacity for eliminating both ionic and nonionic parts, ED has been utilized in the chemical and food industries as well as to treat saltwater and industrial wastes. The variety of ED technological uses is actually the result of the development of IEM. For the purpose of producing irrigation water, it is possible to employ monovalent perm-selective IEM to selectively separate ions that are monovalent from multivalent ions. Additionally, a lot of research has produced and improved mathematical models as a potent instrument to design ED system to assess and improve its efficacy. Additionally, reverse-ED (RED) and photovoltaic (PV) technologies were combined with ED technique to support greener DS, reduce the consumption of energy from fossil fuels, and reduce the production of millions of tons of carbon dioxide. However, traditional ED powered by the electrical grid is still less expensive than ED-PV, which raises the cost of the drinkable water generated by renewable-ED. Additionally, there are still issues with concentration diffusion, electromigration, proton leak from the AEM, and IEM fouling. As a result, just 4% of the world's DS capability is built, and most ED advancements continue to occur at the laboratory level (Al-Amshawee et al., 2020).

7.2.2 Electrodeionization in desalination

A beneficial technological solution for brackish water's spectrum of salinities is EDI. Although modeling can aid in understanding how EDI cells behave, no trustworthy models linking important operational factors have yet to be established (Otero et al., 2021). As a result of its ability to eliminate a variety of pollutants, EDI offers adaptable options for separations in the biological, chemical, food, and medicinal sectors. Like any modern technologically advanced product, EDI is continually changing, with every new version bringing advances in deionization efficiency and higher single module flow rate capabilities. The primary technical factors that affect an EDI module's performance are temperature, TDS, current strength, and flow rates in the DLC and concentrate compartment (CC). Less complexities, more dependability, smaller footsteps, and cheaper components and labor costs are all past developments in EDI systems. The primary operating costs for EDI installations are associated with EDI module replacement, consumption of electricity, wastewater treatment, and laboratory monitoring. The system performance and the feed water concentration have a direct impact on these expenses (Arar et al., 2014). Fig. 7.2 shows the schematic representation for DS using EDI.

Electrodeionization in desalination and water softening 167

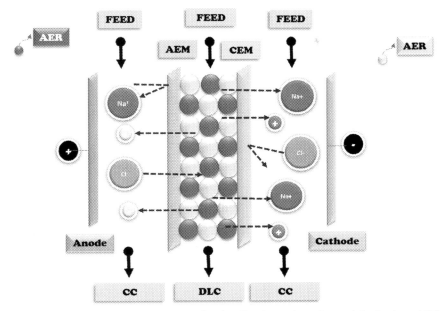

Figure 7.2 Schematic representation for desalination using electrodeionization. *AEM*, Anion exchange membrane; *AER*, anion exchange resin; *CC*, concentrate compartment; *CEM*, cation exchange membrane; *CER*, cation exchange resin; *DLC*, dilute compartment.

In order to produce water for drinking from seawater, EDI reversal (EDIR), a type of EDI employing regular shifts of polarity, was investigated. The conductance, TDS, and quantities of the main anions or cations in the desalted water that is produced tend to be smaller than the accepted limits established by the Chinese and the World Health Organization standards for the quality of drinking water over an extensive range of feed TDS concentrations (between 2000 and 4000 ppm) in artificial seawater. It was demonstrated that utilizing suitable voltages being applied and PR period ranges, the EDIR system may be utilized to create water for drinking from brine water that has high TDS concentrations up to 4000 ppm without scale development. This investigation suggests a viable method for producing water to drink for some regions, particularly those where a lack of freshwater is a serious problem and when the only possibilities for drinkable water are brackish and/or saltwater DS (Sun et al., 2016).

Because of the loose IER in the EDI stack, the conventional EDI systems have two drawbacks: the irregular distribution of flow in the stack, which results in unsteady and reduced effectiveness of removal, and the leakage of ions within the sections, which has an important effect on the effectiveness of removal and expenses of the DS method (Pan et al., 2017). As a result, a comprehensive strategy for addressing these difficulties has been suggested, known as resin wafer EDI (RW-EDI), which involves immobilizing the loose beads of resin into a wafer shape. RW-EDI is simpler to construct and operates more effectively than standard EDI. It is possible to improve RW-EDI technique by utilizing various IERs and IEMs. The ideal working parameters were 2.28 V per cell pair for the voltage that was used and 810 mL/min for the feed rate, which translated to a production rate of 55.5 L/h m^2 and consumption of energy of 0.66 kWh/m^3. RW-EDI was the saltwater DS technique with the highest energy efficiency when compared to other types of water DS methods (Zheng et al., 2018).

The membrane-free EDI (MFEDI) has performed well to clean the artificially acidic RO permeate of tap water. This device is loaded exclusively with mixed strong-acid and strong-base resins. After the MFEDI device was given a current density greater than its critical current density, 75 A/m^2, water dissociation reaction might happen on the interface surface of the CER and AER. Effective renewal and good cleansing were accomplished. After DS, the artificial RO permeate's conductance dropped from its initial value of 10 S/cm to between 0.060 and 0.062 S/cm. The worn-out IER may restart DS after they had recovered. Water recuperation and energy use were 0.35−0.41 kW h/m^3 water and 93.1%, correspondingly. The mixed concentrate gathered during renewal, which has a pH value of 6.5 and an average conductance of 145 S/cm, might be sent back to a pretreatment unit, such RO, for recovery. There was not a requirement for any chemicals, and no effluent was created (Hu et al., 2015).

7.3 Water softening

Most naturally occurring, man-made, and residential waters include carbonate and calcium ions, which can cause major failures in setups by causing scaling issues. Scaling can be avoided utilizing a number of techniques, such as acidifying the water, employing IER, MTs, or promoting the precipitation of calcium carbonate by a chemical reaction. Another method involves using chemical inhibitors to stop crystal formation and nucleation. However, the use of these compounds in water for drinking is prohibited since they are often harmful to people's well-being (Gabrielli et al., 2006). Precipitation by chemicals has been employed extensively to soften water on a municipal scale in alongside IE. This method of WS, sometimes referred to as lime softening, precipitates the ions of calcium as calcium carbonate and magnesium ions are formed as magnesium hydroxide by adding lime to mineral-rich water. The creation of a large amount of lime slurry and the necessity of using reagents such quick lime, coagulating agents, sodium bicarbonate, and acids for pH adjustment are disadvantages of lime softening. Other techniques for softening water comprise NF, ED, carbon nanotubes, and RO, although they all require large amounts of energy and can be expensive to operate and maintain. Consequently, there is a critical need to create an energy-efficient, economical, and ecologically sound technique that would be a perfect complement to the softening of water industry (Brastad and He 2013).

7.3.1 Different techniques for water softening

The total calcium and magnesium levels affect how hard the water is for drinking. The procedure of softening involves lowering the water's hardness. The main benefits of softening "hard" water for drinking are reductions in (1) copper and lead soluble content, (2) the quantity of chemical cleaners required, and (3) the amount of calcium carbonate scaling in water-heating appliances, which results in a higher degree of ease for people (Beeftink et al., 2021). CP and IE are the two most often utilized techniques for WS. In the CP technique, bicarbonates were transformed into carbonates as precipitates and the pH of the water was raised by the addition of alkaline additions. The restricted solubility of calcium and magnesium carbonates makes it impossible to entirely eliminate the hardness of water with this technique. IERs are especially suited for getting rid of ionic contaminants that can cause issues in production, steam generation, and a variety of industrial and commercial uses. In IE procedures where cations are extracted from solution, common polymeric IERs are typically used. It should be noted that if inexpensive IERs, such zeolites, are employed, the IE is seen as being cost-effective (Entezari and Tahmasbi 2009). Although MTs such as RO, NF, and ED are accessible for this purpose, the costs associated with producing the concentrates must still be kept to a minimum in order to increase the technology's competitiveness. Table 7.2 represents the efficiency of WS with different

Table 7.2 Efficiency of water softening (WS) with different techniques along with its operating condition.

Salt removal	Technique	Operating condition	Efficiency for WS (%)	References
Calcium carbonate	Electrocoagulation process and electrochemical precipitation	Initial concentration: 300 ppm Shaker speed: 200 rpm Current density: 34.63 A/m^2 Flow rate: 116.43 mL/min	95	Zhi and Zhang (2014)
Calcium	Electrodeionization	Initial concentration: 4.94 mg/dm^3 Flow rate: 6 dm^3/h Voltage: 10 V Time: 6 h	100	Fu et al. (2009)
Calcium	Electrodeionization	Initial concentration: 50 ppm Flow rate: 9 mL/min Current: 19.0 mA Time: 4 h pH: 7	96.69	Zhang and Chen (2016)
Magnesium	Electrodeionization	Initial concentration: 50 ppm Flow rate: 9 mL/min Current: 19.0 mA Time: 4 h pH: 7	96.13	Zhang and Chen (2016)
Calcium	Electrodeionization reversal	Initial concentration: 250 ppm Flow rate: 200 mL/min Time: 2 h pH: 4	99.8	Lee et al. (2012)
Magnesium	Electrodeionization reversal	Initial concentration: 250 ppm Flow rate: 200 mL/min Time: 2 h pH: 4	99.6	Lee et al. (2012)
Calcium	Ion exchange	Initial concentration: 230 ppm Shaker speed: 200 rpm Time: 12 h Flow rate: 3.33 mL/min	93	Altundoğan et al. (2016)

Magnesium	Ion exchange	Initial concentration: 155 ppm Shaker speed: 200 rpm Time: 12 h Flow rate: 3.33 mL/min	79	Altundoğan et al. (2016)
Calcium, magnesium	Ion exchange	Initial concentration: 240 ppm Shaker speed: 200 rpm Time: 24 h Flow rate: 14 mL/min	100	Bibiano-Cruz et al. (2016)
Calcium	Nanofiltration	Initial concentration: 158 ppm Pressure: 30 bar Temperature: 15°C	70	Schaep et al. (1998)
Magnesium	Nanofiltration	Initial concentration: 10.1 ppm Pressure: 30 bar Temperature: 15°C	70	Schaep et al. (1998)

techniques along with its operating condition. The electrochemical WS technique is a fascinating alternative to the various methods that have been used to address the water's hardness problems (Sanjuán et al., 2019).

7.3.1.1 Ion exchange

The basis for IE is the phenomena of identically charged ions swapping among a medium and a solution of electrolytes. A complicated cross-linked polymer matrix often used as an IER can serve as the media. For the manufacture of ultrapure water as well as WS, IERs have been frequently employed. The advantages of IER include its ease of use and the absence of energy requirements for the exchange phenomena, while its drawbacks include its limited resin exchange capacity and chemical use during renewal (Al Abdulgader et al., 2013). The numerous uses for IE methods and IER, particularly in the industrial sector, as well as the growing popularity of these innovations in the domestic/residential sector, show this method's capacity for water treatment to partially or completely eradicate the magnesium and calcium content. The expansion of the product portfolios of the major production businesses as well as the rise of newer companies contributed to the IER substance market's increased competitiveness, which put pressure on manufacturers to lower their prices. The engineer choosing the IER substance for his particular process is now faced with a broad range of substances and pricing as well as a dearth of scientific records, which might result in poor choice of material (Lazar et al., 2014).

7.3.1.2 Reverse osmosis

The treatment and disposal of RO concentrates are significant concerns to take into account, even if RO is an attractive option for wastewater recovery. Traditional chemical and physical treatment techniques for RO concentrate have certain drawbacks, including the need for an additional step to remove hardness and generally inefficient phosphorus and nitrogen elimination rates (Wang et al., 2016). To test the viability of the electrocoagulation technique as a potential replacement for the traditional preliminary treatments employed to minimize fouling of membranes before treating DS and WS by RO method, such as chemical coagulation, chlorine treatment, and scale inhibiting agents, researchers utilized it as a substitute preprocessing (Hakizimana et al., 2016).

Because RO guarantees an excellent rejection rate of minerals and colloidal particles, it is frequently coupled with additional pretreatment approaches for the superior treatment and advantageous reuse of generated water. The water that is generated must be prepared because contaminants harm RO membranes through the two processes called clogging and scaling. Membrane fouling is largely brought on by the pore-blocking effects of colloidal material on the membrane outer layer or the adsorption of fat along with other fine organic matter. Scaling often causes the permeate flow to be impeded, the drop in pressure to rise, pore clogging to become permanent, and physical harm to the membrane. Therefore, while operating the majority of RO methods, scale prevention is a crucial factor. It has been hypothesized that adding coagulants before RO can address these issues by raising the standard of the output water and lowering the fouling of the membrane. However, there are several drawbacks to chemical the coagulation process including the need for more space, significant chemical usage, significant sludge production, and insufficient clean-up of light and very small oily particles. Specifically, for cleaning wastewater that is oily with light dispersed oil particles, the electrocoagulation process has been recognized as a promising approach. Additionally, there are research showing that electrocoagulation can reduce hardness and stop the build-up of scale. Because of this, when cleaning specific types of water, electrocoagulation is likely a more efficient RO preprocessing method than chemical coagulation methods (Zhao et al., 2014).

7.3.1.3 Nanofiltration

Pesticides, hardness, and nitrates may all be partially or concurrently eliminated by a method called NF, allowing for a single-step cleaning. To enable effective management of the membrane-based unit, it is necessary to investigate the elimination qualities in the NF stage (Van der Bruggen et al., 2001). NF is a very efficient method for eliminating hardness. With the membrane, multivalent ion retention rates exceeded 90%, while monovalent ion retention rates were between 60% and 70%. For calcium, a rejection rate of 94% was discovered, and it remained constant at greater yields. One equation that links the permeate flow to water viscosity and net pressure variation might adequately capture the impact of temperature and recuperation on the permeate flux. It was discovered that a little amount of organic molecules might result in a significant decrease in flow (Schaep et al., 1998).

The NF procedure offers the advantages of simplicity of use, dependability, and very low consumption of energy in addition to extremely effective pollution elimination. This reduces the likelihood of scale building up on the RO and thermal DS process-related equipment. As a result, NF membranes have drawn interest on a global scale. Recently, both thermal and membrane seawater DS procedures have used NF membranes in preliminary processing unit operations. This has led to less chemicals being required in prior treatment procedures as well as lower energy and water production expenses, which have resulted in more ecologically friendly procedures (Izadpanah and Javidnia 2012). Conventional NF methods have not been able to effectively eliminate a range of pollutants while reducing the incidence of fouling of membranes because to the rising complexity of feed water concentration. It is a biological contact oxidized precipitating ultrafiltration technique, which incorporates improved coagulation, physical adsorption, biological oxidation, and depth treatment by the membrane-based method as a preliminary process. It successfully treats high-hardness micropolluted water and produces effluent with good quality that satisfies "Sanitary Standard for Drinking Water" standards (Wang et al., 2020).

7.3.1.4 Electrodialysis

The basic idea behind ED is to use IEM to remove ions from water-based solutions while being propelled by an electric field. The voltage provided to the membrane cells controls how much DS is taking place, and with it, how much nitrate and hardness are eliminated. The volume of water used for dilution throughout the concentrate cycle determines the concentrate's grade (Hell et al., 1998). The application of an enormous potential resulted in a brief period of operation. On the opposite hand, it was less clear how flow rate affected the efficiency of separation. Additionally, it was discovered that divalent particles operate more slowly than monovalent particles do. The potential that was applied was shown to have an impact on the consumption of energy. Energy usage, nevertheless, was less susceptible to changes in flow rate (Kabay et al., 2002).

With the right current settings, one may achieve excellent elimination of salt in this technique. Nevertheless, in the CC of an electrodialyzer, problematic sediments could form on the outermost layer of the membrane in the case of high DS of water with a high concentration of calcium and magnesium ions. Scaling is a phenomenon that impairs the characteristics of the membrane, raises the system's resistance, and raises the amount of energy required for ED. By switching the electrode's polarity or by adding the right reagents to the CC stream, scaling's impacts can be lessened. Scaling can also be avoided by swapping out problematic ions for neutrality ones that, even at large concentrations, are incapable of creating deposits with counter-ions (Wiśniewski and Różańska 2007).

7.3.1.5 Adsorption

For the absorption of diverse ionic and molecular forms from water, adsorption has been extensively researched. While activated carbon is an effective adsorbent, its cost continues to be a problem. Therefore, discovering low-cost and effective adsorbents continues to be a hot subject for the water treatment sectors. Pumice stone is a kind of volcanic rock that results from the intense eruption of highly pressurized, superheated rock, which solidifies foamy lava. Lava that has been heated and water mixed together may also make pumice. It is thought that concurrent fast cooling and depressurization caused this unusual structure. Utilizing natural and alkali-modified pumices as adsorbents, it was explored how to soften hard water by eliminating ions such as calcium and magnesium. Cations were removed more effectively when the adsorbent weight, contact duration, or starting ions level were increased. When contrasted to magnesium, the investigated pumice adsorbents demonstrated a greater selectivity for calcium adsorption (Sepehr et al., 2013).

Adsorptive membrane is a viable choice for cleaning water since it combines the adsorption and MT into one system. The most straightforward and extensively used approach for creating adsorptive membranes is the coupling of solid adsorbent materials with a matrix of polymers (Qiao et al., 2022). Research have been done into modifying zeolite to be utilized as an adsorbent by adding titanium dioxide. As a result of the newly developed active functional group's existence and improved zeolite's chemical properties, it can be said that the zeolite's overall efficacy toward WS has improved overall, especially at lower pH levels where titanium dioxide is added to the zeolite. Additionally, the alteration did not impact the primary structure of zeolite and cuts the ideal equilibrium period required for treating ground water by 29%−50%. Furthermore, while employing modest adsorbent doses and the ideal pH range, the percentage of hardness eliminated through zeolite and titanium dioxide modified zeolite increased significantly (Mubarak et al., 2022).

7.3.2 Electrodeionization in water softening

The EDI system demonstrated an outstanding elimination rate for all ions while handling real samples of ground water at minimal and steady cell

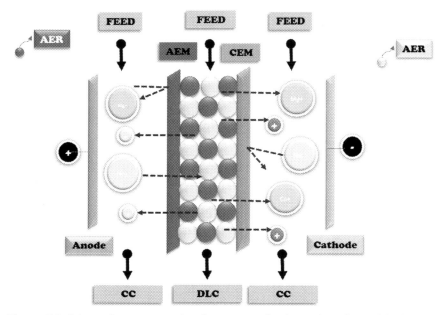

Figure 7.3 Schematic representation for water softening using electrodeionization. *AEM*, anion exchange membrane; *AER*, anion exchange resin; *CC*, concentrate compartment; *CEM*, cation exchange membrane; *CER*, cation exchange resin; *DLC*, dilute compartment.

voltage. The findings supported the EDI's suitability for the constant and extremely effective elimination and recuperation of hardness ions from ground water. The integration of IER among the membranes of water that are enclosed inside the DLC of the ED system allows systems that feature EDI to continually treat liquids with low conductance. The applied electric field, which may dissolve water on-site into hydrogen ions and hydroxide, constantly regenerates the IER, negating the need for additional chemicals. Additionally, additional IERs offer additional conductance to the DLC, resulting in a more quick, efficient, and thorough dissociation (Zhang and Chen 2016). Fig. 7.3 shows the schematic representation for WS using EDI.

It was discovered that using the ideal voltage for operation range might yield a calcium elimination rate of almost 100%. The risk of stack scaling may significantly rise if the CC rate of flow is inadequate. Even though the calcium and magnesium ions in the IER bed might interchange, the experiment of the multispecies feed mixture did not demonstrate any discernible effects on the proportion of total hardness removed.

The elimination of low amounts of hardness ions from aqueous solutions via the EDI technique was suggested to be possible under ideal circumstances, and this is expected to happen in the HPW manufacturing process in the near future (Fu et al., 2009).

To avoid or reduce corrosion of membranes and scaling during the functioning of electromembrane techniques, passion for PR has developed. In the EDIR method with an EDI with the PR, Lee et al. (2012) evaluated the impact of the operational circumstances on the process performance. Divalent cation elimination effectiveness (magnesium and calcium) remained high without surface membrane-scale development. The process efficiency was impacted by PR periods during the EDIR procedure, and the operational PR duration was decided to be 40 minutes or lower. The homogeneity membranes show somewhat greater efficiency in eliminating than the heterogeneity membrane in the for a long-time EDIR studies with various CER. Using electron microscopy to analyze the scaling powder that had developed on the outermost layer of the membrane, it was discovered that the majority of the scale had developed as the carbonate form of calcium and magnesium. The viability of the EDIR apparatus for usage in the WS technique has been effectively proven by Lee et al., (2012) without scale development in a suitable concentration limit.

7.4 Conclusion

A beneficial technique is DS and WS, which has the ability to turn what is believed to be essentially endless ocean water reserves into drinking water. Nevertheless, the same water-saving techniques that are promoted today must still be used, particularly because DS plants' ability for output is still a constraint. Furthermore, in developing countries that will be badly impacted by the consequences of climate change and water shortages, it is crucial to identify drinkable water sources that can be more economical than traditional DS facilities. The numerous techniques for DS and WS, as well as their applications, were covered in this chapter. The application of EDI in DS and WS has also undergone recent research. EDI surpasses other techniques in terms of cost and environmental consideration. As a result of recent technology developments, DS is now more effective and affordable on a worldwide level. This has been made feasible by advances in the materials employed in MT-based DS, the introduction of energy recovery technologies to lower power requirements, and the synthesis of several DS techniques into hybrid systems. In addition, alternative sources of energy have been gradually phased into power DS facilities, which will aid in ensuring the company's long-term viability.

References

Abdel-Rehim, Z.S., Lasheen, A., 2005. Improving the performance of solar desalination systems. Renewable Energy 30 (13), 1955−1971.

Al Abdulgader, H., Kochkodan, V., Hilal, N., 2013. Hybrid ion exchange−pressure driven membrane processes in water treatment: a review. Separation and Purification Technology 116, 253−264.

Al-Amshawee, S., Yunus, M.Y.B.M., Azoddein, A.A.M., Hassell, D.G., Dakhil, I.H., Hasan, H.A., 2020. Electrodialysis desalination for water and wastewater: a review. Chemical Engineering Journal 380. Available from: https://doi.org/10.1016/j.cej.2019.122231.

Alkaisi, A., Mossad, R., Sharifian-Barforoush, A., 2017. A review of the water desalination systems integrated with renewable energy. Energy Procedia 110, 268−274.

Alklaibi, A.M., Lior, N., 2005. Membrane-distillation desalination: status and potential. Desalination 171 (2), 111−131.

Altundoğan, H.S., Topdemir, A., Çakmak, M., Bahar, N., 2016. Hardness removal from waters by using citric acid modified pine cone. Journal of the Taiwan Institute of Chemical Engineers 58, 219−225. Available from: https://doi.org/10.1016/j.jtice.2015.07.002, http://www.elsevier.com/wps/find/journaldescription.cws_home/715607/description#description.

Antar, M.A., Bilton, A., Blanco, J., Zaragoza, G., 2012. Solar desalination. Annual review of heat transfer.

Arar, O., Yüksel, U., Kabay, N., Yüksel, M., 2014. Various applications of electrodeionization (EDI) method for water treatment—a short review. Desalination 342, 16−22. Available from: https://doi.org/10.1016/j.desal.2014.01.028.

Arjunan, T.V., Aybar, H.Ş., Nedunchezhian, N., 2009. Status of solar desalination in India. Renewable and Sustainable Energy Reviews 13 (9), 2408−2418.

Banasiak, L.J., Kruttschnitt, T.W., Schäfer, A.I., 2007. Desalination using electrodialysis as a function of voltage and salt concentration. Desalination 205 (1−3), 38−46.

Beeftink,, M., Hofs,, B., Kramer,, O., Odegard, I., Van Der Wal, A., 2021. Carbon footprint of drinking water softening as determined by life cycle assessment. Journal of Cleaner Production 278. Available from: https://doi.org/10.1016/j.jclepro.2020.123925.

Bibiano-Cruz, L., Garfias, J., Salas-García, J., Martel, R., Llanos, H., 2016. Batch and column test analyses for hardness removal using natural and homoionic clinoptilolite: breakthrough experiments and modeling. Sustainable Water Resources Management 2 (2), 183−197. Available from: https://doi.org/10.1007/s40899-016-0050-y, springer.com/journal/40899.

Boyd, C.E., Tucker, C.S., Somridhivej, B., 2016. Alkalinity and hardness: critical but elusive concepts in aquaculture. Journal of the World Aquaculture Society 47 (1), 6−41.

Brastad, K.S., He, Z., 2013. Water softening using microbial desalination cell technology. Desalination 309, 32−37.

Campanile, A., Liguori, B., Ferone, C., Caputo, D., Aprea, P., 2022. Zeolite-based monoliths for water softening by ion exchange/precipitation process. Scientific Reports 12 (1). Available from: https://doi.org/10.1038/s41598-022-07679-2.

Cohen, Y., Semiat, R., Rahardianto, A., 2017. A perspective on reverse osmosis water desalination: quest for sustainability. AIChE Journal 63 (6), 1771–1784.

Curto, D., Franzitta, V., Guercio, A., 2021. A review of the water desalination technologies. Applied Sciences 11 (2), 670.

Entezari, M.H., Tahmasbi, M., 2009. Water softening by combination of ultrasound and ion exchange. Ultrasonics Sonochemistry 16 (3), 356–360.

Fu, L., Wang, J., Su, Y., 2009. Removal of low concentrations of hardness ions from aqueous solutions using electrodeionization process. Separation and Purification Technology 68 (3), 390–396.

Gabrielli, C., Maurin, G., Francy-Chausson, H., Thery, P., Tran, T.T.M., Tlili, M., 2006. Electrochemical water softening: principle and application. Desalination 201 (1–3), 150–163. Available from: https://doi.org/10.1016/j.desal.2006.02.012.

Ghaffour, N., Missimer, T.M., Amy, G.L., 2013. Technical review and evaluation of the economics of water desalination: current and future challenges for better water supply sustainability. Desalination 309, 197–207.

Goh, P.S., Ismail, A.F., 2018. A review on inorganic membranes for desalination and wastewater treatment. Desalination 434, 60–80.

Hakizimana, J.N., Gourich, B., Vial, Ch, Drogui, P., Oumani, A., Naja, J., et al., 2016. Assessment of hardness, microorganism and organic matter removal from seawater by electrocoagulation as a pretreatment of desalination by reverse osmosis. Desalination 393, 90–101. Available from: https://doi.org/10.1016/j.desal.2015.12.025.

Hawlader, M.N.A., Ho, J.C., Teng, C.K., 2000. Desalination of seawater: an experiment with RO membranes. Desalination 132 (1–3), 275–280.

Hell, F., Lahnsteiner, J., Frischherz, H., Baumgartner, G., 1998. Experience with full-scale electrodialysis for nitrate and hardness removal. Desalination 117 (1–3), 173–180. Available from: https://doi.org/10.1016/s0011-9164(98)00088-5.

Hu, J., Fang, Z., Jiang, X., Li, T., Chen, X., 2015. Membrane-free electrodeionization using strong-type resins for high purity water production. Separation and Purification Technology 144, 90–96. Available from: https://doi.org/10.1016/j.seppur.2015.02.023, http://www.journals.elsevier.com/separation-and-purification-technology/.

Izadpanah, A.A., Javidnia, A., 2012. The ability of a nanofiltration membrane to remove hardness and ions from diluted seawater. Water 4 (2), 283–294.

Kabay, N., Demircioglu, M., Ersöz, E., Kurucaovali, I., 2002. Removal of calcium and magnesium hardness by electrodialysis. Desalination 149 (1–3), 343–349. Available from: https://doi.org/10.1016/s0011-9164(02)00807-x.

Kalogirou, S., 1997. Survey of solar desalination systems and system selection. Energy 22 (1), 69–81.

Karagiannis, I.C., Soldatos, P.G., 2008. Water desalination cost literature: review and assessment. Desalination 223 (1–3), 448–456.

Laycock, M.V., Anderson, D.M., Naar, J., Goodman, A., Easy, D.J., Donovan, M.A., et al., 2012. Laboratory desalination experiments with some algal toxins. Desalination 293, 1–6. Available from: https://doi.org/10.1016/j.desal.2012.02.014.

Lazar, L., Bandrabur, B., Tataru-Fărmuş, R.E., Drobotă, M., Bulgariu, L., Gutt, G., 2014. FTIR analysis of ion exchange resins with application in permanent hard water softening. Environmental Engineering and Management Journal 13 (9), 2145–2152. Available from: https://doi.org/10.30638/eemj.2014.237.

Lee, H.J., Hong, M.K., Moon, S.H., 2012. A feasibility study on water softening by electrodeionization with the periodic polarity change. Desalination 284, 221–227.

Li, H., Zou, L., 2011. Ion-exchange membrane capacitive deionization: a new strategy for brackish water desalination. Desalination 275 (1−3), 62−66.

Liu, L., Cheng, Q., 2020. Mass transfer characteristic research on electrodialysis for desalination and regeneration of solution: a comprehensive review. Renewable and Sustainable Energy Reviews 134, 110115.

Liu, Q., Xu, G.R., Das, R., 2019. Inorganic scaling in reverse osmosis (RO) desalination: mechanisms, monitoring, and inhibition strategies. Desalination 468, 114065.

Misdan, N., Lau, W.J., Ismail, A.F., 2012. Seawater reverse osmosis (SWRO) desalination by thin-film composite membrane—current development, challenges and future prospects. Desalination 287, 228−237.

Mubarak, M.F., Mohamed, A.M.G., Keshawy, M., Elmoghny, T.A., Shehata, N., 2022. Adsorption of heavy metals and hardness ions from groundwater onto modified zeolite: batch and column studies. Alexandria Engineering Journal 61 (6), 4189−4207. Available from: https://doi.org/10.1016/j.aej.2021.09.041, http://www.elsevier.com/wps/find/journaldescription.cws_home/724292/description#description.

Otero, C., Urbina, A., Rivero, E.P., Rodríguez, F.A., 2021. Desalination of brackish water by electrodeionization: experimental study and mathematical modeling. Desalination 504. Available from: https://doi.org/10.1016/j.desal.2020.114803.

Pan, S.Y., Snyder, S.W., Ma, H.W., Lin, Y.J., Chiang, P.C., 2017. Development of a resin wafer electrodeionization process for impaired water desalination with high energy efficiency and productivity. ACS Sustainable Chemistry and Engineering 5 (4), 2942−2948. Available from: https://doi.org/10.1021/acssuschemeng.6b02455, http://pubs.acs.org/journal/ascecg.

Pan, S.Y., Snyder, S.W., Ma, H.W., Lin, Y.J., Chiang, P.C., 2018. Energy-efficient resin wafer electrodeionization for impaired water reclamation. Journal of Cleaner Production 174, 1464−1474. Available from: https://doi.org/10.1016/j.jclepro.2017.11.068.

Panagopoulos, A., Haralambous, K.J., 2020. Environmental impacts of desalination and brine treatment-Challenges and mitigation measures. Marine Pollution Bulletin 161, 111773.

Pugsley, A., Zacharopoulos, A., Mondol, J.D., Smyth, M., 2016. Global applicability of solar desalination. Renewable Energy 88, 200−219. Available from: https://doi.org/10.1016/j.renene.2015.11.017, http://www.journals.elsevier.com/renewable-and-sustainable-energy-reviews/.

Qiao, L., Ye, H., Xin, Q., Huang, L., Zhang, Y., Li, H., 2022. An adsorptive sulfonated polyethersulfone/functionalized graphene ultrafiltration membrane for hardness removal. Separation and Purification Technology 287. Available from: https://doi.org/10.1016/j.seppur.2022.120567.

Qiblawey, H.M., Banat, F., 2008. Solar thermal desalination technologies. Desalination 220 (1−3), 633−644.

Sadrzadeh, M., Mohammadi, T., 2008. Sea water desalination using electrodialysis. Desalination 221 (1−3), 440−447.

Sagle, A.C., Van Wagner, E.M., Ju, H., McCloskey, B.D., Freeman, B.D., Sharma, M.M., 2009. PEG-coated reverse osmosis membranes: desalination properties and fouling resistance. Journal of Membrane Science 340 (1−2), 92−108. Available from: https://doi.org/10.1016/j.memsci.2009.05.013.

Sanjuán, I., Benavente, D., Expósito, E., Montiel, V., 2019. Electrochemical water softening: influence of water composition on the precipitation behaviour. Separation and Purification Technology 211, 857−865. Available from: https://doi.org/10.1016/j.seppur.2018.10.044, http://www.journals.elsevier.com/separation-and-purification-technology/.

Schaep, J., Van Der Bruggen, B., Uytterhoeven, S., Croux, R., Vandecasteele, C., Wilms, D., et al., 1998. Removal of hardness from groundwater by nanofiltration. Desalination 119 (1−3), 295−301. Available from: https://doi.org/10.1016/S0011-9164(98)00172-6, https://www.journals.elsevier.com/desalination.

Seo, S.J., Jeon, H., Lee, J.K., Kim, G.Y., Park, D., Nojima, H., et al., 2010. Investigation on removal of hardness ions by capacitive deionization (CDI) for water softening applications. Water Research 44 (7), 2267–2275. Available from: https://doi.org/10.1016/j.watres.2009.10.020.

Sepehr, M.N., Zarrabi, M., Kazemian, H., Amrane, A., Yaghmaian, K., Ghaffari, H.R., 2013. Removal of hardness agents, calcium and magnesium, by natural and alkaline modified pumice stones in single and binary systems. Applied Surface Science 274, 295–305. Available from: https://doi.org/10.1016/j.apsusc.2013.03.042, http://www.journals.elsevier.com/applied-surface-science/.

Sevda, S., Yuan, H., He, Z., Abu-Reesh, I.M., 2015. Microbial desalination cells as a versatile technology: functions, optimization and prospective. Desalination 371, 9–17. Available from: https://doi.org/10.1016/j.desal.2015.05.021.

Shahmirzadi, M.A.A., Hosseini, S.S., Luo, J., Ortiz, I., 2018. Significance, evolution and recent advances in adsorption technology, materials and processes for desalination, water softening and salt removal. Journal of Environmental Management 215, 324–344.

Shen, X., Li, T., Jiang, X., Chen, X., 2014. Desalination of water with high conductivity using membrane-free electrodeionization. Separation and Purification Technology 128, 39–44. Available from: https://doi.org/10.1016/j.seppur.2014.03.011.

Singh, K., Zhang, L., Zuilhof, H., de Smet, L.C.P.M., 2020. Water desalination with nickel hexacyanoferrate electrodes in capacitive deionization: Experiment, model and comparison with carbon. Desalination 496. Available from: https://doi.org/10.1016/j.desal.2020.114647.

Strathmann, H., 2004. December. Assessment of electrodialysis water desalination process costs. In Proceedings of the International Conference on Desalination Costing, Limassol, Cyprus (pp. 32–54).

Subramani, A., Jacangelo, J.G., 2015. Emerging desalination technologies for water treatment: a critical review. Water Research 75, 164–187. Available from: https://doi.org/10.1016/j.watres.2015.02.032.

Sun, X., Lu, H., Wang, J., 2016. Brackish water desalination using electrodeionization reversal. Chemical Engineering and Processing: Process Intensification 104, 262–270.

Tai, Z.S., Abd Aziz, M.H., Othman, M.H.D., Mohamed Dzahir, M.I.H., Hashim, N.A., Koo, K.N., et al., 2020. Ceramic membrane distillation for desalination. Separation and Purification Reviews 49 (4), 317–356. Available from: https://doi.org/10.1080/15422119.2019.1610975, http://www.tandfonline.com/toc/lspr20/current.

Tao, S., Xu, Y.D., Gu, J.Q., Abadikhah, H., Wang, J.W., Xu, X., 2018. Preparation of high-efficiency ceramic planar membrane and its application for water desalination. Journal of Advanced Ceramics 7 (2), 117–123. Available from: https://doi.org/10.1007/s40145-018-0263-7, http://www.springer.com/materials/special + types/journal/40145.

Turek, M., 2003. Cost effective electrodialytic seawater desalination. Desalination 153 (1–3), 371–376.

Valavala, R., Sohn, J.S., Han, J.H., Her, N.G., Yoon, Y.M., 2011. Pretreatment in reverse osmosis seawater desalination: a short review. Environmental Engineering Research 16 (4), 205–212.

Van der Bruggen, B., Everaert, K., Wilms, D., Vandecasteele, C., 2001. Application of nanofiltration for removal of pesticides, nitrate and hardness from ground water: rejection properties and economic evaluation. Journal of Membrane Science 193 (2), 239–248. Available from: https://doi.org/10.1016/s0376-7388(01)00517-8.

Villafafila, A., Mujtaba, I.M., 2003. Fresh water by reverse osmosis based desalination: simulation and optimisation. Desalination 155 (1), 1–13.

Wang, J., Wang, S., Jin, M., 2000. A study of the electrodeionization process—high-purity water production with a RO/EDI system. Desalination 132 (1–3), 349–352.

Wang, X.X., Wu, Y.H., Zhang, T.Y., Xu, X.Q., Dao, G.H., Hu, H.Y., 2016. Simultaneous nitrogen, phosphorous, and hardness removal from reverse osmosis concentrate by microalgae cultivation. Water Research 94, 215−224. Available from: https://doi.org/10.1016/j.watres.2016.02.062.

Wang, Y., Ju, L., Xu, F., Tian, L., Jia, R., Song, W., et al., 2020. Effect of a nanofiltration combined process on the treatment of high-hardness and micropolluted water. Environmental Research 182. Available from: https://doi.org/10.1016/j.envres.2019.109063.

Warsinger, D.M., Swaminathan, J., Guillen-Burrieza, E., Arafat, H.A., Lienhard V, J.H., 2015. Scaling and fouling in membrane distillation for desalination applications: a review. Desalination 356, 294−313. Available from: https://doi.org/10.1016/j.desal.2014.06.031.

Wiśniewski, J., Różańska, A., 2007. Donnan dialysis for hardness removal from water before electrodialytic desalination. Desalination 212 (1−3), 251−260.

Ye, G., Yu, Z., Li, Y., Li, L., Song, L., Gu, L., et al., 2019. Efficient treatment of brine wastewater through a flow-through technology integrating desalination and photocatalysis. Water Research 157, 134−144. Available from: https://doi.org/10.1016/j.watres.2019.03.058.

Younos, T., 2005. Environmental issues of desalination. Journal of Contemporary Water Research and Education 132 (1), 3.

Younos, T., Tulou, K.E., 2005. Overview of desalination techniques. Journal of Contemporary Water Research & Education 132 (1), 3−10.

Zhang, Z., Chen, A., 2016. Simultaneous removal of nitrate and hardness ions from groundwater using electrodeionization. Separation and Purification Technology 164, 107−113.

Zhao, S., Huang, G., Cheng, G., Wang, Y., Fu, H., 2014. Hardness, COD and turbidity removals from produced water by electrocoagulation pretreatment prior to reverse osmosis membranes. Desalination 344, 454−462. Available from: https://doi.org/10.1016/j.desal.2014.04.014.

Zheng, X.Y., Pan, S.Y., Tseng, P.C., Zheng, H.L., Chiang, P.C., 2018. Optimization of resin wafer electrodeionization for brackish water desalination. Separation and Purification Technology 194, 346−354. Available from: https://doi.org/10.1016/j.seppur.2017.11.061, http://www.journals.elsevier.com/separation-and-purification-technology/.

Zhi, S., Zhang, S., 2014. A novel combined electrochemical system for hardness removal. Desalination 349, 68−72.

CHAPTER EIGHT

Production of high pure water using electrodeionization

8.1 Introduction

Ultrapure water (UPW) is water that has had impurities purposefully reduced to exceedingly minute levels. A part of the worldwide water marketplace is taken up by UPW, which is used for boiler/cooling water in power manufacturing as well as for high-value-added sectors including semiconductors (SCs) and drug manufacturing. Over the past few years, UPW has become more and more in need. As a result of meeting the needs of numerous industrial uses, UPW production today holds a place in the global water market that is practically on line with saltwater desalination (DS). The average UPW manufacturing system comprises 5—10 purifying techniques since an exceptionally low level of pollutants is required (Lee et al., 2016). Nuclear and thermal power technology is the largest user of UPW. High levels of water purification reduce the danger of apparatus deterioration and the accumulation of deposits on it, improving reliability of operations and the cost-effective functioning of nuclear, thermal, and combined heat and power plants. Making SC and microelectronics is the second-biggest industry that uses UPW. The very sensitive examination of tiny levels of components is currently another significant application area for UPW (Melnik and Krysenko, 2019).

The UPW manufacturing method is a very intricate procedure that makes use of several purifying techniques. The standard UPW manufacturing system primarily consists of three phases: pretreatment, DS, and polishing are the first three stages. Every procedure will be impacted by the results of the preceding stage due to the accumulated nature of the whole production chain. In order to guarantee the wastewater quality, each stage in the UPW manufacturing process is crucial. Traditionally, adsorption, RO, UV irradiation, and other methods have been utilized to remove organic components from the manufacturing of UPW. The two-pass RO procedure may successfully lower the level of organic chemicals in UPW among them (Zhang et al., 2021). UPW production in nuclear power plants (NPPs) and technological industries needs electrodeionization (EDI). In contrast to the conventional ion-exchange (IE) method, EDI is not required for a regenerative process of high-pollution resin with chemical reagents. But since the process uses a lot of energy, cutting back on energy use is essential. By improving the operating conditions and unit design, the decrease may be accomplished. Consequently, EDI is an eco-friendly method for producing UPW. The complete membrane method is quickly replacing IE methods in the generation of UPW in order to eliminate the harmful effects of the wastewater treatment process (Qian et al., 2022). This chapter aims to provide a comprehensive review of UPW, its properties, and its applications such as SC industries, power generation, nanotechnology, and pharmaceutical industries (PIs). Also, it explores the UPW production techniques such as reverse osmosis (RO), ultrafiltration, ion exchanger, and EDI.

8.2 Ultrapure water

The efficiency and efficacy of high-tech items, such as SCs, electronics, and medicines, all of which need extremely precise manufacture, are greatly influenced by UPW. Additionally, the utilization of UPW is crucial for enhancing the energy effectiveness of NPPs and boiler installations that generate a significant volume of steam. Therefore, as these sectors develop, so does the need for UPW in different procedures. As a result, creative methods that effectively manufacture UPW are emerging (Lee and Choi, 2012). In order to eliminate organic substances and inorganics, UPW manufacturing includes preprocessing and refining phases. However, creating UPW free of contamination from bacteria is still one of the largest issues confronting UPW companies. Almost every part of the UPW method can be impacted by bacteria once they are present. These effects include a rise in biofouling of RO membranes, frictional and heat transfer resistance, accumulation of contaminants on filtration systems, and colonization of pipes, which ultimately results in damage to the supply line (McAlister et al., 2002).

UPW is water that has been purified using many methods to get river water or water intended for use in industry as near to its original purity as feasible. In addition to the particles and minerals, the UPW also removes

Production of high pure water using electrodeionization 185

gas that has been dissolved in the water. UPW is free of any pollutants or contaminants that could be found in regular drinking freshwater. Where the water comes from might affect the kinds of pollutants and their concentration that exist in your water. Data may suffer greatly if there are pollutants and contaminants present. By making sure the water you use is extremely pure, you can prevent data tampering and guarantee accurate, dependable results (Ahuja, 2009). The properties of UPW are shown in Table 8.1

Table 8.1 Properties of ultrapure water.

Property	Quantity	References
Ammonium	50 pg/mL	Ehmann et al. (2006)
Boron	0.05 μg/dm^3	Melnik and Krysenko (2019)
Bromide	20 pg/mL	Ehmann et al. (2006)
Calcium	2 pg/mL	Ehmann et al. (2006)
Chloride	20 pg/mL	Ehmann et al. (2006)
Chromium	<0.02 μg/L	Chen (1997)
Conductivity	0.08 μS/cm at 25°C	Mamba et al. (2009)
Copper	<0.1 μg/L	Chen (1997)
Dissolved oxygen	3 μg/dm^3	Melnik and Krysenko (2019)
Fluoride	30 pg/mL	Ehmann et al. (2006)
Iron	<0.1 μg/L	Chen (1997)
Lithium	3 pg/mL	Ehmann et al. (2006)
Magnesium	2 pg/mL	Ehmann et al. (2006)
Manganese	<0.05 μg/L	Chen (1997)
Maximum micro organisms	1 CFU/mL	Bennett (2009)
Maximum total dissolved solids	0.005 ppm	Bennett (2009)
Maximum total organic carbon	0.005 ppm	Bennett (2009)
Nitrate	20 pg/mL	Ehmann et al. (2006)
Nitrite	20 pg/mL	Ehmann et al. (2006)
Phosphate	20 pg/mL	Ehmann et al. (2006)
Potassium	5 pg/mL	Ehmann et al. (2006)
Reactive silica	2 μg/L	Bennett (2009)
Resistivity	18 MΩ cm at 25°C	Bennett (2009)
Sodium	5 pg/mL	Ehmann et al. (2006)
Sulfate	20 pg/mL	Ehmann et al. (2006)
Turbidity	<0.1 FTU	Mamba et al. (2009)
Zinc	< 0.1 μg/L	Chen (1997)

The removal of all components other than molecules of water is the primary goal of UPW manufacture. In reality, this is practically unachievable, the purity grade fluctuates according to the particular requirements of a certain procedure. For example, because UPW is utilized for injecting water in the PI, it is of the utmost importance that it be devoid of microorganisms and endogenous pyrogens. Managing rust and scaling is essential for water cooling systems used in a variety of production techniques, particularly NPPs, as the process of exchanging heat quickly increases precipitation of chemicals, that can completely ruin the system (Greenlee et al., 2009). Due to the highly sensitive nature of the accuracy type equipment, the SC market in particular requires the greatest degree of purity. An electrical resistivity of a minimum of 18 MΩ cm (at 25°C) and very low pollution concentrations are necessary. The microelectronic level UPW system thereby captures the core of UPW technology by utilizing the most intricate and extensive manufacturing line (Lee, 2005).

8.3 Application of ultrapure water

The word "UPW" refers to very pure water, as the name implies; however, even under this general phrase, many UPW-using businesses have varying requirements and purity demands. For example, the power-generating, medicinal, and SC industries (SCIs) all use UPW in their respective fields of business (Hughes et al., 1971). The criteria for each sector's purity requirements differ; for example, the medical device sector's top priorities are "reproducibility, accuracy, and documentation," while both the power generation and SCI demand extremely low electrical conductivity and silica levels. The SCI has the most exacting standards of all the sectors adopting UPW (Fitzsimons, 2011).

8.3.1 Power generation

In UPW, which is characterized by incredibly low salinity and mineral levels, bacteria can induce biofouling and biocorrosion, thus reducing lifespan and raising expenses for operation. Sector including SC, medicine, food and beverage manufacturing, heating and cooling systems, and NPPs are just a few that are affected by the microbes that are part of UPW. The majority of the time, UPW is utilized as feed, cooling water, or raw material. The study of their bacterial diversity has achieved international relevance since UPW is an essential part of many industrial type surroundings (Bohus et al., 2010).

The need to keep plants running as long as feasible while keeping them safe and economically viable is at the cutting edge of the power challenge. It is necessary to maintain control over the deteriorating of components, materials, and constructions in order to complete this task. Since the 1970s, there have been several corrosion-related breakdowns at nuclear power facilities, costing businesses trillions of dollars (Cattant et al., 2008). In order to manage heat in NPPs, the main refrigerant is a crucial cooling media. UPW is the principal coolant that is utilized the most frequently. The main circuits' surfaces discharge contaminants from corrosion into the cooling system when a NPP is operating. These substances of corrosion may get into the nuclear reactor's center and become active, which would cause activity to form on the main system's surfaces, the system and its parts to become contaminated, and workers at NPP to be exposed to radiation while the plant is shut down. In order to lower the amount of radiation in a NPP, system purification is thus a crucial aspect (Yeon et al., 2004).

The maintenance of UPW, the foundation for the production of UPW steam in an NPP water-steam cycle, is one of the main problems in NPP. Raw water, boiler feed water, demineralizer influent/effluent, boiler blowdown water, process steam, high- and low-pressure steam condensate, and condensate polisher water should all be checked for cations, anions, heavy metals, and silica. For the detection and avoidance of corrosive situations in various NPP components, the monitoring of trace levels of ionic contaminants across the power production process is essential. Reduced and ongoing monitoring is required for ions that are corrosive like chloride. Controlling contaminants, including ions of sodium, sulfate ions, and various other ions (phosphate, fluoride, calcium, nitrate, and magnesium), can give significant details about the origin of damage, the likelihood that contaminants will accumulate upwards, the probability that corrosion will occur, as well as accurate data for NPP start-up and closure (Cickaric et al., 2005).

8.3.2 Semiconductor industries

SCIs are a vital component of our lives. Any piece of electronic gadgetry that is utilized on a daily basis, from little items such as earphones to larger ones such as cell phones, computers, and car controls, cannot operate without a SC (Hasan and Yu, 2017). The manufacture of SC is a technology-intensive as well as a highly energy-intensive enterprise. In 1996, SCI utilized almost 8 billion kWh. A SCI's electrical power consumption is significantly impacted by the quantity of UPW and pure gases utilized in and provided by operations that demand a lot of electrical power (Hu and Chuah, 2003). Because to UPW's microbiological pollution, several businesses are affected. The SCI is one of these. One cell of bacteria or the byproducts of cellular disintegration might seriously impair the overall quality of the end item since the SCI uses UPW in the last rinse cycle.

The preliminary treatment and polishing phases are two of the main steps in the multistep, complicated industrial manufacture of UPW. Many UPW manufacturing processes contain a number of stages to eliminate and eradicate microorganisms. Particularly, microbiological contamination is avoided in specific areas of a facility by ozonation and UV254 light therapy. To stop CO_2 and O_2 from dissolving in the water, nitrogen gas is frequently utilized rather than air above stored UPW. In order to avoid ionic loading on the mixed-bed IE resins (MIERs), it is essential that UPW remains CO_2-free, and reducing the oxygen content should limit development of bacteria (Kulakov et al., 2002).

SCI uses a lot of water for a range of tasks, including as washing wafer surfaces and cooling machinery (Tsai et al., 2002). Many stages of the process call for UPW. Wafer washing, cleaning, and surface conditioning must be done repeatedly throughout the first phases of device manufacture. It is utilized for surface cleaning, wet etching, solvent being processed, and chemical mechanical planarization at several phases of the device production process. Due to the enormous quantities needed for slurry formation and rinsing, the latter unit operation has in fact grown to be one of the fab's biggest users of UPW (Boyd and Dornfeld, 2012). Therefore, all SCIs engage in a major and expensive effort to transform raw water into UPW. There are ongoing and major attempts within the manufacturing sector to decrease the consumption of UPW due to the expensive price of manufacture and the huge volume requirements.

8.3.3 Pharmaceutical industries

A crucial part of many procedures, particularly those widely employed in the PI, is UPW. UPW must only include water and equal concentrations of

H + and OH − ions in order to be classified (Jallouli et al., 2018). Within the SCI and PI, UPW is used as a crucial reagent in a variety of laboratory environments. Although these two sectors are where UPW is most frequently seen, it may be used for any form of extremely sensitive laboratory work due to its high degree of purification (Shetty and Goyal, 2022).

The PI represents one of the areas where UPW is most frequently used. High degrees of filtration are essential to ensure the quality of goods in this industry as it handles medical products (Baz et al., 1991). According to the essential requirements, water used in PI has to have dissolved solids concentrations at least 10,000 times lower than that of regular drinking water (Belkacem et al., 2008). Regular potable water fails to meet these requirements. Water utilized in this sector must also be absolutely free of microorganisms and organic matter. In total, 98% of the minerals in the influent water must be eliminated using the RO method in order to lower the amount of suspended matter in this water. After this first stage, subsequent treatments are used to generate water that is as pure as is required (Khan et al., 2023) by employing sophisticated IE or EDI systems.

However, UPW may include pollutants even if it claims a high degree of resistivity. Since organic pollutants, endotoxins, and nucleases have no effect on resistivity values, it is frequently necessary to use alternate methods to get rid of these impurities. It is common to call this kind of purifying apparatus a "polisher." These polishers are an essential component in the creation of UPW and can be either fed by a localized RO unit or a centralized ring main. In the PI, the quality of UPW is very crucial. It is used as a beginning product or solvent for the creation of injection-based solutions because of its special qualities (Schäfer et al., 2022).

8.4 Ultrapure water production

A number of purifying methods, including RO, IE, UV radiation, EDI, different kinds of filtration, and degasification, are used to produce UPW. The basic purification techniques utilized to create municipal water—which serves as the feed water to the UPW plant, are the first two unit processes (Khajavi et al., 2010). Fig. 8.1 represents the UPW production plant.

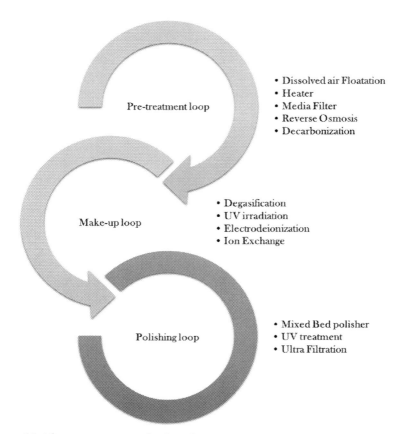

Figure 8.1 Ultrapure water production plant.

This municipal water will typically go through three steps of filtration, with several sequential unit procedures included in each level. A preliminary processing or make-up process, a main process, and an end polishing process normally comprise the three looping processes (Fitzsimons, 2011). Table 8.2 gives the various techniques for UPW production and its properties.

8.4.1 Ultrafiltration

A complex series of purification processes are used to produce UPW, and they may be generally grouped into the following categories: (1) a preliminary treatment phase that eliminates suspended matter, (2) a make-up phase to desalt, and (3) a cleaning phase to increase the water's quality to a level

Table 8.2 Various techniques for ultrapure water (UPW) production and its properties.

Technique with specification	Initial conductivity/ concentration ($\mu s/cm$)	Product conductivity ($\mu s/cm$)	pH of UPW	References
EDI 3 compart ments Bipolar membrane IER: IRN77	100	7.7	6	Yeon et al. (2003)
EDI 3 compartments Bipolar membrane IER: SKN1	100	4.2	6	Yeon et al. (2003)
EDI 3 compartments Bipolar membrane IER: IR120	100	6.5	6	Yeon et al. (2003)
EDI Membrane free EDI Two reticular electrodes	217	0.059	7.9	Su et al. (2013)
RO/EDI 5 compartments Length of chamber: 30 cm Width of the chamber: 10 cm	400	20	7	Wang et al. (2000)
RO/EDI 5 compartments Length of chamber: 22 cm Width of the chamber: 31 cm	58.17	0.4	6.3	Wenten and Arfianto (2013)

(Continued)

Table 8.2 (Continued)

Technique with specification	Initial conductivity/concentration (µs/cm)	Product conductivity (µs/cm)	pH of UPW	References
RO/EDI 5 compartments mixed ion exchange resin (MER)	5	0.05	6	Ervan and Wenten (2002)
RO/EDI Capacity: 50 gpm	3.54	0.067	—	Prato and Gallagher (2010)
RO/EDI/IE Capacity: 800 gpm	3.4	0.057	—	Prato and Gallagher (2010)
Ultrafiltration (UF)/EDI 5 compartments MER DLC: 8 mm	248	1	7.1	Wardani et al. (2017)
UF/RO/EDI Capacity: 360 gpm	4.56	0.0625	—	Prato and Gallagher (2010)

EDI, Electrodeionization; *IE*, ion exchange; *IER*, ion-exchange resin; *RO*, reverse osmosis.

high enough to fulfill the demands of the manufacturing processes (Zhan et al., 2020). In the "UF" separating process, a solution comprising a solute whose molecular dimensions is substantially larger than those of the solvents are stripped of their solute by being compelled by a hydraulic gradient of pressure to pass via an appropriate membrane. Normal filtration, "RO," and UF simply vary in the dimension of the particles that are separated; any other difference between each of them is mostly arbitrary. The majority of what ensues deals with aquatic systems since UF techniques are now mostly restricted to aqueous medium (Michaels et al., 1971).

Novel separation opportunities are provided by the UF method. Numerous and diverse commercial uses for UF may be discovered in a variety of sectors (Toledano et al., 2010). The chemical, electrical, culinary, and biotechnological sectors all employ UF for recovering products and pollution management. In comparison with thermal separation techniques, membrane methods use less energy, which makes UF an appealing concentration option in numerous fields. Despite the UF method uses significantly fewer resources than evaporation does, a big UF plant has a high initial investment cost. As a result, UF can be used as a substitute to evaporation in small- and medium-sized plants, where the available evaporation capacity is constrained, or when handling heat-sensitive goods. The low temperatures of operation of UF enable the treatment of delicate biological substances without causing chemical or physical harm to the contents. According to the study of Jönsson and Trägrdh (1990), the primary issue with UF is the flow reduction brought on by concentration polarization and fouling.

The flux drop from chemically reversible fouling to hydrodynamically reversible fouling appears to be redistributed by coagulation preprocessing. Both the unprocessed and the pretreated waters have almost the same amount of total flow after surface and cleaning with chemicals. When flocculation circumstances led to particles with a zeta potential that was close to zero, coagulation preprocessing inhibited the reduction in membrane flow the greatest. This is assumed to be caused by variations in cake permeability and size of particles rather than better foulant elimination. The biggest decrease in organic matter across the membrane's surface and the highest filtrate output were achieved by preprocessing with ferric chloride under circumstances that led to particles with a zeta potential close to zero (Lahoussine-Turcaud et al., 1990).

The most modern last filtration stage before the UPW supply loop is UF using permeable polymeric excellent membranes (Czuba et al., 2021). These membranes often have two skins and are asymmetrical. The UPW generated using the two procedures combined will have the maximum level of particle cleanliness possible. As a result, this will work best to provide the ground-work for free of flaw production in all phases of the process when UPW is used. For the production of SC devices, UF is used to filter out debris, colloids, and big macromolecules from UPW. This ultimate objective and the initial stage toward free of flaws production need to identify the earliest signs of water leaks, tiny holes, and fiber breakage in membranes units. Monitoring of particles alone is insufficient. In addition to being an essential supplemental tool for preventative care, regular pressure-hold testing allows for substantially higher accuracy (Ruth and Berndt, 2016).

8.4.2 Reverse osmosis

The separation-recovery technique known as RO has been used in a variety of separation-recovery procedures (Marchetti et al., 2014). The capacity to minimize process issues brought on by clogging of the membrane's outer layer as well as substantial advancements in the production of thin-film membranes made of composites have led to a constantly growing market-place for RO procedures (Ikeda et al., 1994). Since then, advancements in new-generation membrane technologies that retain better water flow and solute separating properties under more demanding operating circumstances have pushed RO to the leading edge of separation techniques (Lee and Kim, 2015). A significantly wider range of applications has resulted from the development of hybridized techniques that integrate RO with other traditional separation methods such as UF, NF, distillation, or crystallization. In addition to the conventional saltwater and saltwater DS processes, RO membranes are used for a variety of different applications, including waste-water treatment, the creation of UPW, WS, processing of food, pharmaceutical recovery, and many more (Bhattacharyya et al., 2000).

As they make it possible to remove the majority of impurities from source water, RO membranes are becoming more and more recognized as a basic technology of UPW production systems (Pandey et al., 2012). In contrast to EDI methods, nevertheless, relatively little is understood about the RO membranes utilized for UPW synthesis. There has not been any extensive investigation on the fouling properties of RO membranes that are used for UPW production, despite a recent study emphasizing the relevance of RO membranes in UPW production systems in light of the most current developments in the global water market (Rho et al., 2019).

Production of high pure water using electrodeionization   195

To concurrently create UPW and drinking water from saltwater, the combination of RO and CDI has been researched (Minhas et al., 2014). When the feed intensity was 25,000, 35,000, or 40,000 ppm, accordingly, the UPW was found to have a purity of 1.567, 1.916, and 2.156 ppm. In overall, all three examples had drinkable water concentrations of below 400 ppm, what the WHO considers to be of satisfactory quality. When saltwater is present in concentrations of 25,000, 35,000, and 40,000 ppm, accordingly, the energy required to produce UPW and drinkable water from it is 3.171, 4.195, and 4.565 kWh/m^3. In places where UPW from saltwater is necessary, the suggested combined RO-CDI arrangement is thus of great importance (Jande et al., 2015).

8.4.3 Ion-exchange process

An IE reaction is a balanced exchange of ions among a solid phase and a liquid phase that is reversible, the IE is often insoluble in the solution that is used for the exchange of ions (Pashkov et al., 2016). IE is a very beneficial addition to other processes such as filtration, the process of distillation and adsorption nowadays and is widely recognized as a unit operation. In order to produce UPW, IE was often used as an affordable and practical separation technique. Therefore, IE is employed to supply a significant amount of UPW to PI, food, and SCI. IE is employed in a variety of chemical reactions that fall under the three major categories of ion replacement, ion segregation, and ion elimination. In ion segregation, heavy-metal recovery, hydrometallurgy, drinking-water treatment, and catalytic processes, IE is very useful (Pandey, 2008).

MIER is a key technique for decreasing ionic contaminants to concentrations below pg/dm3 in the continuous manufacturing of UPW on a large scale. It is kinetics instead of the equilibrium conditions of the IE mechanisms that becomes the key governing element under these demanding working circumstances, notably the high flow rate and large volume needs of condensate purifying in the power-generating sector (Harries, 1991). The major objective of the IE procedure, which comes after RO, is to create 18.2 MΩ cm resistivity by eliminating any remaining anions and cations in solution. Even while organic compounds having the charge-negative functional groups could be readily eliminated by the IER (Choi et al., 2016), the role of IE in the removal of carbon from organic matter in water is not fully proven.

By radiation graft glycidyl methacrylate onto a polyethylene nonwoven material and then chemically modifying it, IE fabric (IEF) with the functional group of sulfonic acid was created. After rinsing with organic solvents, the amount of all organic carbon that had been eluted from the resultant IEF could be decreased to a level lower than 1 ppb. The resulting IEF's adsorption capacity was assessed using a 10 ppb sodium solution. With a distribution coefficient of 1.2 $\times 10^7$ and a linear speed of 400 m/h, the 7-mm diameter and 20-mm height column loaded IEF was able to remove sodium. At 95-mm column height, the tipping point in bed capacity was attained at 2.0×10^5, and column utilization increased to 18.7%. Based on these findings, the IEF produced by graft polymerization was determined to be suitable for water filtration in the manufacture of UPW (Takeda et al., 2010).

8.4.4 Electrodeionization

A method for ionic isolation called EDI first appeared around half a century ago. It was employed to eradicate metallic ions from radioactive wastewater in its initial use; nevertheless, a lack of knowledge of its functional characteristics has hindered its growth and usage (Rathi et al., 2022). Investigations have been steadily expanding and have been centered on the clarification of complicated operating mechanisms, allowing for the expansion of its applicability to various disciplines (Wen et al., 2005). EDI has so far been shown to be a great, ecologically friendly technique for concentration, segregation, and cleansing. In order to advance and mature this field of study, novel substances have been continually created (Alvarado and Chen, 2014). If successful, this sort of technology might have a significant positive impact on the environment and the worldwide economy (Widiasa et al., 2004).

Electrodialysis (ED) and IE technologies are used in the hybrid separating technique referred to as EDI. As in the case of an ED system, cation exchange membranes (CEMs) and anion exchange membranes (AEMs) are positioned across the electrodes in the system (Alvarado et al., 2009). A direct current is used to drive the IER in the center chamber, which improves the transfer of cations or anions. Hydrogen (H +) and hydroxyl (OH) ions are produced during EDI operation by the water breakdown process. Without the need of regeneration chemicals, generated hydrogen ions and hydroxyl ions constantly renew the IER electrochemically. The permselective AEM and CEM are alternated in an EDI device. The membrane gaps are designed to form compartments for liquid flow with inlets and outputs. Electrodes are placed at the extremities of the IEM and chambers and are powered by an external power source (Rathi and Kumar, 2020).

Due to the IEM's permselective characteristics and the potential voltage gradient's orientation, ions that exist in the water solution have diminished in the DLC and concentrated in the nearby CC. The process's fundamental component is the DLC's utilization of IER. IE is not feasible for the majority of sources of clean water without the conductivity of ions provided by the IER, which is one explanation for this. There is a more insignificant, although equally significant, electrochemical link that exists between the applied voltage and the equilibrium amount of hydrogen ions and hydroxyl ions in water, which is the second justification for the requirement for IER in the DLC. A sizable number of H+ and OH ions are created at concentrated locations with strong potential gradients. The drawbacks of localized pH variations and ineffective current utilization result from the creation of hydrogen ions and hydroxyl ions without the inclusion of IER (Arar et al., 2014).

UPW production in high-tech and NPP sectors is impossible without EDI. In contrast to the conventional IE procedure, EDI lacks a requirement for the regenerative process of the high-pollution IER by chemical reagents. But because the process uses a lot of energy, cutting back on the consumption of energy is essential. Operating conditions and design of the modules can be optimized to accomplish a reduction. The energy effectiveness of ion elimination and hence EDI optimization depend heavily on the water dissociation processes; however, they are rarely taken into account (Qian et al., 2022). Fig. 8.2 represents the EDI unit for UPW production. The EDI technique now fills a comparatively tiny void

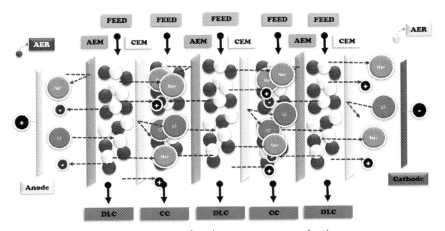

Figure 8.2 Electrodeionization unit for ultrapure water production.

in the field of water treatment industry, refining RO permeates to create UPW. Making the process adaptable of a larger variety of feed solutions conditions, maybe permitting easing of the feed criteria and enabling more uses of EDI, could be a possible area of improvement that might be pursued in the years to come. The price of EDI module has significantly decreased over the past few years as a result of advancements in modules design, such the usage of thick cells. However, there is still a sizable window of possibility to further cut the price of EDI systems by optimizing the skid's pipelines, electrical supply, and controls designs, which are yet to be optimized (Wood et al., 2010).

Both a rise in applied voltage and a rise in flow rate of water result in an improvement in UPW impedance. Interpolymer-based IEM at 25 volts/cell pair applied voltage reached UPW of impedance 18.2 MΩ m with product rate 18.5 L/h, according to standardization trials. Current efficiency (CE) and power consumption (W) were measured at 0.324 kWh/kg and 58.86%, correspondingly. UPW was resistant to Ionsep and Fujifilm type II commercial membranes by 15 and 17.1 MΩ cm, respectively, at an applied voltage of 30 V/cell pair. Ionsep membranes had W and CE values of 0.658 kWh/kg and 34.83%, accordingly, whereas Fujifilm type II membranes had values of 0.430 kWh/kg and 53.88%. Ionsep membranes provide UPW with a reduced resistance because to their rapid water absorption and low transport number compared to interpolymer-based IEM (Bhadja et al., 2015).

Over time, EDI keeps producing water that is dependably of a high quality. In addition, conductance and particular ion rejection efficiency is extremely high, foreseeable, and steady over time. Additionally, weakly ionized molecules such as carbon dioxide and silica may be reliably and consistently removed by EDI units (Lu et al., 2010). EDI differs from other enormous-scale industrial membrane processes in that it can reliably and predictably eliminate dissolved ionic species in addition to poorly ionized species without adding chemicals to the input stream. The operational performance of these machines illustrates the EDI method's commercial potential for UPW manufacturing facilities of this size. Particularly when combined with RO, EDI produces water of high purity. The purity level attained is often at or close to the UPW standard specification. The benefits of EDI over traditional IE systems are considerable: By working in a continuous mode, EDI is naturally more reliable and less susceptible to the disturbances caused by batch processes (Hernon et al., 1994). EDI eliminates the requirement for on-site storage, utilization, and ultimately elimination of huge quantities of regenerating chemicals.

8.5 Conclusion

This chapter investigated the comprehensive overview of UPW, its properties, and its applications such as SC industries, power generation, and PIs. Also, it has been explored the UPW production techniques such as RO, ultrafiltration, ion exchanger, and EDI. UPW is essential in numerous commercial uses, including generating electricity, SCI, and PI. The IER beads improve the transfer of mass, make splitting water easier, and lower stack impedance in EDI, which is employed to create UPW. The generation of UPW for the PI, power, and SC sectors is presently being done by EDI at several facilities all over the world. Over the coming few years, it is anticipated that EDI will play a larger role in the manufacture of UPW. When compared to the cost of maintaining the current IE beds, RO/EDI proved an economically viable alternative. Additional research is required to better understand the ideal hydrodynamic parameters of the IER bed, IER contamination, and IER stability over time.

References

Ahuja, S. (Ed.), 2009. Handbook of Water Purity and Quality. Academic press.

Alvarado, L., Chen, A., 2014. Electrodeionization: principles, strategies and applications. Electrochimica Acta 132, 583−597.

Alvarado, L., Ramírez, A., Rodríguez-Torres, I., 2009. Cr (VI) removal by continuous electrodeionization: study of its basic technologies. Desalination 249 (1), 423−428.

Arar, O., Yüksel, U., Kabay, N., Yüksel, M., 2014. Various applications of electrodeionization (EDI) method for water treatment—a short review. Desalination 342, 16−22. Available from: https://doi.org/10.1016/j.desal.2014.01.028.

Baz, M., Durand, C., Ragon, A., Jaber, K., Andrieu, D., Merzouk, T., et al., 1991. Using ultrapure water in hemodialysis delays carpal tunnel syndrome. The International Journal of Artificial Organs 14 (11), 681−685. Available from: https://doi.org/10.1177/039139889101401101.

Belkacem, M., Bensadok, K., Refes, A., Charvier, P.M., Nezzal, G., 2008. Water produce for pharmaceutical industry: role of reverse osmosis stage. Desalination 221 (1−3), 298−302. Available from: https://doi.org/10.1016/j.desal.2007.01.086.

Bennett, A., 2009. Water processes and production: high and ultra-high purity water. Filtration & Separation 46 (2), 24−27.

Bhadja, V., Makwana, B.S., Maiti, S., Sharma, S., Chatterjee, U., 2015. Comparative efficacy study of different types of ion exchange membranes for production of ultrapure water via electrodeionization. Industrial and Engineering Chemistry Research 54 (44), 10974−10982. Available from: https://doi.org/10.1021/acs.iecr.5b03043. Available from: http://pubs.acs.org/journal/iecred.

Bhattacharyya, D., Mangum, W.C., Williams, M.E., 2000. Reverse osmosis. Kirk-Othmer Encyclopedia of Chemical Technology.

Bohus, V., Tóth, E.M., Székely, A.J., Makk, J., Baranyi, K., Patek, G., et al., 2010. Microbiological investigation of an industrial ultra pure supply water plant using cultivation-based and cultivation-independent methods. Water Research 44 (20), 6124–6132. Available from: https://doi.org/10.1016/j.watres.2010.07.006.

Boyd, S., Dornfeld, D., 2012. Semiconductor manufacturing. Green Manufacturing: Fundamentals and Applications. Springer US, Boston, MA, pp. 153–178.

Cattant, F., Crusset, D., Féron, D., 2008. Corrosion issues in nuclear industry today. Materials Today 11 (10), 32–37.

Chen, G., 1997. Measurement and control of organic contaminants in ultrapure water systems. The University of Arizona.

Choi, J., Kim, J.O., Chung, J., 2016. Removal of isopropyl alcohol and methanol in ultrapure water production system using a 185 nm ultraviolet and ion exchange system. Chemosphere 156, 341–346.

Cickaric, D., Dersek-Timotic, I., Onjia, A., Rajakovic, L., 2005. Development of ion chromatography methods for the determination of trace anions in ultra pure water from power plants. Journal of the Serbian Chemical Society 70 (7), 995–1003. Available from: https://doi.org/10.2298/jsc0507995c.

Czuba K., Bastrzyk A., Rogowska A., Janiak K., Pacyna K., Kossińska N., et al., 2021. Towards the circular economy—a pilot-scale membrane technology for the recovery of water and nutrients from secondary effluent. Science of The Total Environment.

Ehmann, T., Mantler, C., Jensen, D., Neufang, R., 2006. Monitoring the quality of ultrapure water in the semiconductor industry by online ion chromatography. Microchimica Acta 154, 15–20.

Ervan, Y., Wenten, I.G., 2002. Study on the influence of applied voltage and feed concentration on the performance of electrodeionization. Songklanakarin Journal of Science and Technology 24, 955–963.

Fitzsimons, L., 2011. A detailed study of desalination exergy models and their application to a semi-conductor ultra-pure water plant (Doctoral dissertation, Dublin City University).

Greenlee, L.F., Lawler, D.F., Freeman, B.D., Marrot, B., Moulin, P., 2009. Reverse osmosis desalination: Water sources, technology, and today's challenges. Water Research 43 (9), 2317–2348. Available from: https://doi.org/10.1016/j.watres.2009.03.010.

Harries, R.R., 1991. Ion exchange kinetics in ultra-pure water systems. Journal of Chemical Technology & Biotechnology 51 (4), 437–447.

Hasan, M.S., Yu, H., 2017. Innovative developments in HCI and future trends. International Journal of Automation and Computing 14, 10–20.

Hernon B.P. Zanapalidou R.H. Zhang L. Siwak L.R. Schoepke E.J. 19941994 Proceedings of 55th Annual Meeting International Water Conference Application of electrodeionization in ultrapure water production: performance and theory.

Hu, S.C., Chuah, Y.K., 2003. Power consumption of semiconductor fabs in Taiwan. Energy 28 (8), 895–907.

Hughes, R.C., Mürau, P.C., Gundersen, G., 1971. Ultrapure water. preparation and quality. Analytical Chemistry 43 (6), 691–696.

Ikeda, T., Muragishi, H., Bairinji, R., Uemura, T., 1994. Advanced reverse osmosis membrane modules for novel ultrapure water production process. Desalination 98 (1–3), 391–400. Available from: https://doi.org/10.1016/0011-9164(94)00165-0.

Jallouli, N., Pastrana-Martínez, L.M., Ribeiro, A.R., Moreira, N.F.F., Faria, J.L., Hentati, O., et al., 2018. Heterogeneous photocatalytic degradation of ibuprofen in ultrapure water, municipal and pharmaceutical industry wastewaters using a TiO2/UV-LED system. Chemical Engineering Journal 334, 976–984. Available from: https://doi.org/10.1016/j.cej.2017.10.045.

Jande, Y.A.C., Minhas, M.B., Kim, W.S., 2015. Ultrapure water from seawater using integrated reverse osmosis-capacitive deionization system. Desalination and Water Treatment 53 (13), 3482−3490.

Jönsson, A.S., Trägårdh, G., 1990. Ultrafiltration applications. Desalination 77, 135−179.

Khajavi, S., Jansen, J.C., Kapteijn, F., 2010. Production of ultra pure water by desalination of seawater using a hydroxy sodalite membrane. Journal of Membrane Science 356 (1−2), 52−57.

Khan, Z.U., Moronshing, M., Shestakova, M., Al-Othman, A., Sillanpää, M., Zhan, Z., et al., 2023. Electro-deionization (EDI) technology for enhanced water treatment and desalination: a review. Desalination 548. Available from: https://doi.org/10.1016/j.desal.2022.116254.

Kulakov, L.A., McAlister, M.B., Ogden, K.L., Larkin, M.J., O'Hanlon, J.F., 2002. Analysis of bacteria contaminating ultrapure water in industrial systems. Applied and Environmental Microbiology 68 (4), 1548−1555. Available from: https://doi.org/10.1128/AEM.68.4.1548-1555.2002. Available from: http://aem.asm.org/.

Lahoussine-Turcaud, V., Wiesner, M.R., Bottero, J.-Y., Mallevialle, J., 1990. Coagulation pretreatment for ultrafiltration of a surface water. Journal - American Water Works Association 82 (12), 76−81. Available from: https://doi.org/10.1002/j.1551-8833.1990.tb07070.x.

Lee, K., 2005. Making a technological catch-up: barriers and opportunities. Asian Journal of Technology Innovation 13 (2), 97−131.

Lee, J.H., Choi, J.H., 2012. The production of ultrapure water by membrane capacitive deionization (MCDI) technology. Journal of Membrane Science 409, 251−256.

Lee, H., Kim, S., 2015. Factors affecting the removal of isopropyl alcohol by reverse osmosis membranes for ultrapure water production. Desalination and Water Treatment 54 (4-5), 916−922.

Lee, H., Jin, Y., Hong, S., 2016. Recent transitions in ultrapure water (UPW) technology: Rising role of reverse osmosis (RO). Desalination 399, 185−197.

Lu, J., Wang, Y.X., Lu, Y.Y., Wang, G.L., Kong, L., Zhu, J., 2010. Numerical simulation of the electrodeionization (EDI) process for producing ultrapure water. Electrochimica Acta 55 (24), 7188−7198. Available from: https://doi.org/10.1016/j.electacta.2010.07.054.

Mamba, B.B., Krause, R.W., Malefetse, T.J., Gericke, G., Sithole, S.P., 2009. Cyclodextrin nanosponges in the removal of organic matter for ultrapure water in power generation. Journal of Water Supply: Research and Technology-Aqua 58 (4), 299−304. Available from: https://doi.org/10.2166/aqua.2009.039.

Marchetti, P., Jimenez Solomon, M.F., Szekely, G., Livingston, A.G., 2014. Molecular separation with organic solvent nanofiltration: A critical review. Chemical Reviews 114 (21), 10735−10806. Available from: https://doi.org/10.1021/cr500006j. Available from: http://pubs.acs.org/journal/chreay.

McAlister, M.B., Kulakov, L.A., O'Hanlon, J.F., Larkin, M.J., Ogden, K.L., 2002. Survival and nutritional requirements of three bacteria isolated from ultrapure water. Journal of Industrial Microbiology and Biotechnology 29 (2), 75−82. Available from: https://doi.org/10.1038/sj.jim.7000273.

Melnik, L.A., Krysenko, D.A., 2019. Ultrapure water: properties, production, and use. Journal of Water Chemistry and Technology 41, 143−150.

Michaels, A.S., Nelsen, L., Porter, M.C., 1971. Ultrafiltration. In Membrane Processes in Industry and Biomedicine: Proceedings of a Symposium held at the 160th National Meeting of the American Chemical Society, under the sponsorship of the Division of Industrial and Engineering Chemistry, Chicago, Illinois, September 16 and 17, 1970 (pp. 197−232). Springer US.

Minhas, M.B., Jande, Y.A.C., Kim, W.S., 2014. Combined reverse osmosis and constant-current operated capacitive deionization system for seawater desalination. Desalination 344, 299−305.

Pandey, A.K., 2008. Kinetic study of ion exchange column operation for ultrapure water application (Doctoral dissertation, Oklahoma State University).

Pandey, S.R., Jegatheesan, V., Baskaran, K., Shu, L., 2012. Fouling in reverse osmosis (RO) membrane in water recovery from secondary effluent: a review. Reviews in Environmental Science and Biotechnology 11 (2), 125−145. Available from: https://doi.org/10.1007/s11157-012-9272-0.

Pashkov, G.L., Saikova, S.V., Panteleeva, M.V., 2016. Reactive ion exchange processes of nonferrous metal leaching and dispersion material synthesis. Theoretical Foundations of Chemical Engineering 50, 575−581.

Prato, T., Gallagher, C., 2010. Using EDI to meet the needs of pure water production. GE Power Water Water Process Technol 1−5.

Qian, F., Lu, J., Gu, D., Li, G., Liu, Y., Rao, P., et al., 2022. Modeling and optimization of electrodeionization process for the energy-saving of ultrapure water production. Journal of Cleaner Production 372. Available from: https://doi.org/10.1016/j.jclepro.2022.133754.

Rathi, B.S., Kumar, P.S., 2020. Electrodeionization theory, mechanism and environmental applications. A review. Environmental Chemistry Letters 18, 1209−1227.

Rathi, B.S., Kumar, P.S., Parthiban, R., 2022. A review on recent advances in electrodeionization for various environmental applications. Chemosphere 289, 133223.

Rho, H., Chon, K., Cho, J., 2019. An autopsy study of a fouled reverse osmosis membrane used for ultrapure water production. Water 11 (6), 1116.

Ruth, J., Berndt, R., 2016, May. Quality control for ultrafiltration of ultrapure water production for high end semiconductor manufacturing. In 2016 27th Annual SEMI Advanced Semiconductor Manufacturing Conference (ASMC) (pp. 16-22). IEEE.

Schäfer, S.H., van Dyk, K., Warmer, J., Schmidt, T.C., Kaul, P., 2022. A new setup for the measurement of total organic carbon in ultrapure water systems. Sensors 22 (5). Available from: https://doi.org/10.3390/s22052004. Available from: https://www.mdpi.com/1424-8220/22/5/2004/pdf.

Shetty, A., Goyal, A., 2022. Total organic carbon analysis in water—a review of current methods. Materials Today: Proceedings.

Su, W., Pan, R., Xiao, Y., Chen, X., 2013. Membrane-free electrodeionization for high purity water production. Desalination 329, 86−92. Available from: https://doi.org/10.1016/j.desal.2013.09.013.

Takeda, T., Tamada, M., Seko, N., Ueki, Y., 2010. Ion exchange fabric synthesized by graft polymerization and its application to ultra-pure water production. Radiation Physics and Chemistry 79 (3), 223−226. Available from: https://doi.org/10.1016/j.radphyschem.2009.08.042.

Toledano, A., García, A., Mondragon, I., Labidi, J., 2010. Lignin separation and fractionation by ultrafiltration. Separation and Purification Technology 71 (1), 38−43. Available from: https://doi.org/10.1016/j.seppur.2009.10.024.

Tsai, W.T., Chen, H.P., Hsien, W.Y., 2002. A review of uses, environmental hazards and recovery/recycle technologies of perfluorocarbons (PFCs) emissions from the semiconductor manufacturing processes. Journal of Loss Prevention in the Process Industries 15 (2), 65−75.

Wang, J., Wang, S., Jin, M., 2000. A study of the electrodeionization process—high-purity water production with a RO/EDI system. Desalination 132 (1-3), 349−352.

Wardani, A.K., Hakim, A.N., Khoiruddin, Wenten, I.G., 2017. Combined ultrafiltration-electrodeionization technique for production of high purity water. Water Science and Technology 75 (12), 2891−2899. Available from: https://doi.org/10.2166/wst.2017.173. Available from: http://wst.iwaponline.com/content/75/12/2891.full.pdf.

Wen, R., Deng, S., Zhang, Y., 2005. The removal of silicon and boron from ultra-pure water by electrodeionization. Desalination 181 (1-3), 153−159.

Wenten, I.G., Arfianto, F., 2013. Bench scale electrodeionization for high pressure boiler feed water. Desalination 314, 109–114.

Widiasa, I.N., Sutrisna, P.D., Wenten, I.G., 2004. Performance of a novel electrodeionization technique during citric acid recovery. Separation and purification technology 39 (1-2), 89–97.

Wood, J., Gifford, J., Arba, J., Shaw, M., 2010. Production of ultrapure water by continuous electrodeionization. Desalination 250 (3), 973–976. Available from: https://doi.org/10.1016/j.desal.2009.09.084.

Yeon, K.H., Seong, J.H., Rengaraj, S., Moon, S.H., 2003. Electrochemical characterization of ion-exchange resin beds and removal of cobalt by electrodeionization for high purity water production. Separation Science and Technology 38 (2), 443–462. Available from: https://doi.org/10.1081/SS-120016584.

Yeon, K.H., Song, J.H., Moon, S.H., 2004. A study on stack configuration of continuous electrodeionization for removal of heavy metal ions from the primary coolant of a nuclear power plant. Water Research 38 (7), 1911–1921.

Zhan, M., Lee, H., Jin, Y., Hong, S., 2020. Application of MFI-UF on an ultrapure water production system to monitor the stable performance of RO process. Desalination 491. Available from: https://doi.org/10.1016/j.desal.2020.114565.

Zhang, X., Yang, Y., Ngo, H.H., Guo, W., Wen, H., Wang, X., et al., 2021. A critical review on challenges and trend of ultrapure water production process. Science of The Total Environment 785. Available from: https://doi.org/10.1016/j.scitotenv.2021.147254.

CHAPTER NINE

Advances future scope in electrodeionization

9.1 Introduction

Ionic method of separation called electrodeionization (EDI) was developed 70 years earlier (Wang et al., 2000). Its growth and utilization have been hindered by a poor understanding of its operational kinetics, which was initially used for eliminating metallic ions from nuclear effluent (Alvarado and Chen, 2014). Both ion-exchange resins (IERs) and ion-exchange membranes (IEMs) are used in the combined separation method known as EDI. The EDI attracted more and more interest for its ability to retrieve or remove ions from water. The EDI may be used in a variety of ways to concentrate and remove various components from wastewater streams (Arar et al., 2014a). The fact that EDI is environmentally friendly is probably the main factor in its financial viability. It removes the leftover stream connected with IER regeneration by replacing the harmful chemicals typically used to renew IER with energy (Wood et al., 2010). The EDI procedure has some distinct advantages over other inorganic solid separation methods, such as increased IER life over ion exchange, a bit longer IEM life over membrane methods such as the reverse osmosis (RO) process and ultrafiltration, simpler IER recuperation, the ability to produce water free of contaminants, and the ability to eliminate and retrieve pure targeted contaminants. This all-natural, eco-friendly detaching strategy has a very promising future and offers a lot of advantages (Rathi and Kumar, 2020).

Although EDI offers a number of noteworthy advantages, it also has a number of disadvantages. Membrane contamination, ion leakage, irregular flow, metal hydroxide scaling, expensive and challenging maintenance are some of these issues (Rathi et al., 2022). The recognized limits of membranes, such as fouling of the membrane, scaling, and concentration polarization, are present in membrane EDI processes. In addition, the formation of metal hydroxides at the IER surface and the anion exchange membranes (AEMs) is unable to be avoided by these procedures (Dermentzis, 2010).

Electrodeionization: Fundamentals, Methods and Applications.
DOI: https://doi.org/10.1016/B978-0-443-18983-8.00009-0
© 2024 Elsevier Inc. All rights are reserved, including those for text and data mining,
AI training, and similar technologies.

Electrostatically shielded EDI (ES-EDI) and membrane-free EDI (MF-EDI) are advancements with innovative arrangement of the EDI membrane stack to avoid the unwanted hydroxide deposition (Dermentzis, 2008). In comparison with present traditional membrane approaches, MF-EDI, ES zone-ionic current sinks (ICS) will explore several electrochemical technical anti-pollution uses and be advantageous in terms of concentration polarization, fouling, and clogging of IEM (Dermentzis et al., 2009b).

High effectiveness in ion elimination occurs in EDI when IERs are positioned in a dilute compartment (DLC) between CEM and AEM. However, when the pH of the solution turns basic during the procedure, scale starts to develop on the surface of IER and IER in a kind of hydroxide compounds, which reduces the overall efficiency of EDI. Scale creation in the procedure can be avoided by the so-called EDI reversal (EDIR) technique, which involves reversing the polarity of the electrodes. Because EDIR can eliminate hard substances without causing significant scale development, as contrasted with equivalent traditional IEM procedures, attention in the periodic shift of polarity has developed (Lee et al., 2013). Due to the inclusion of scattered IER beads, conventional EDI has uneven results and challenging on-site servicing. The speed of ion elimination may be reduced by channel development brought on by loosened IER beads. The stack has to be treated carefully in a controlled manner when there is loosened IER. The free IER beads are being substituted with resin wafers, which are made of the former scattered IER immobilized and molded into a transparent, rigid matrix, to enhance the EDI technique. Ionic mobility is enhanced by resin wafer electrodeionization (RW-EDI), and local pH modulation is also possible (Pan et al., 2017, 2018). The EDI method is examined in this chapter, along with its benefits and drawbacks. Additionally, it looked at the advantages of the most current developments in EDI such as RW-EDI, EDIR, ES-EDI, MF-EDI, and hybrid process along with EDI.

9.2 Electrodeionization technology

EDI is a new and very successful technology for eliminating ionic compounds from water that is contaminated (Widiasa et al., 2004). High-purity water may be produced via the EDI method. Along with producing clean water, the process also has promising methods for wastewater treatment that make it easier to get rid of heavy metals, nuclear waste products, hazardous substances, and other dangerous impurities (Kumar et al., 2022). IERs are continually being renewed by the DC power field as part of the ongoing process known as EDI. The running expenses of EDI are much lower than those of traditional IE methods. A mix of an IER and an IEM are used in the procedure, positioned among two electrodes (anode and cathode) operating at a voltage from a DC supply (Boontawan et al., 2011). The EDI system has lately discovered plenty of novel, innovative uses in the water treatment, the area of biotechnology, and other promising fields. The evolution of stack structure and design are also starting to cause worry, along with continued expansion and expanded applicability (Khoiruddin et al., 2014). Perm-selective AEM and CEM are alternated among anode and cathode in an EDI system. The DLC or product chambers, concentration chambers (CCs), and electrode chambers make up the different sections of an EDI stack. Mixed-bed IER is used to fill the chambers, which improves the movement of ions from bulk solution to the IEM when a direct current is applied. The IER beads improve mass transfer, make the splitting of water easier, and lower stack impedance when EDI is employed to create extremely pure water. Water is divided into hydroxyl and hydrogen ions by the DC electric field, and this process constantly regenerates the IER. The transferred ions are sent to the CC via the membranes and drained out of the system (Wardani et al., 2017).

9.2.1 Advantages of electrodeionization

ED and traditional IE procedures are combined in EDI. The benefit of continuous EDI over mixed bed IER for water ionization is that there is no need for IER renewal, which is often a time-consuming and expensive operation (Ervan and Wenten, 2002). The benefit of continuous EDI over traditional ED is that the electrical conductivity of the IER-filled DLC is improved by over two powers of scale (Wenten and Arfianto, 2013). The EDI method has some unique merits over all other inorganic solid elimination methods, including full recovery of the targeted pollutants, production of water without any pollutants, extra membrane life in comparison with RO, and extra resin lifespan in contrast to IE methods (Gifford et al., 2000). It has been established that EDI is an excellent, environmentally acceptable method for concentrating, isolating, and cleaning. In order to advance and expand this field of knowledge, new compounds are continually being created, which could have significant positive effects on the world economy and ecology (Su et al., 2013).

Compared to traditional IE, chemical renewed deionization process devices, EDI systems are able to provide a number of advantages. Eliminating the regenerating procedure and the dangerous substances, it produces comes first. In contrast to regenerable deionization, where product water purity declines as the resins get closer to depletion, EDI systems maintain a consistent level of water purity as time passes due to the DC power (Zeng et al., 2016). Traditional systems have to be duplexed if constant supply of DI water is necessary in order to ensure a single system may deliver water while another is being regenerated, which increases expense, difficulty, and space. Duplexing is not required because EDI is continual. Consequently, EDI system sizes are frequently half as large as those of their traditional equivalents, and total expenditure on capital for new devices is reduced. This is in addition to the elimination of chemical tanks for storage, chemical pumps, and regenerant waste neutralization apparatus (Strathmann, 2010).

9.2.2 Limitations of existing electrodeionization technology

Despite having many benefits, EDI also has a number of drawbacks. Loosely placed into a pair of AEM and CEM are the IER. Because of convection rather than diffusion, ions escape from a single chamber to another as a result of the loose resin composition that makes section closing difficult. The unbalanced distribution of flow inside the channels, which lowers separating efficacy, is another drawback of loose IER in EDI systems (Ulusoy et al., 2021). Spiral-wound designs have been used in earlier research to solve leakage problems, and they have also been used to solve the channeling issue by immobilizing the IEM via magnetic forces. Only one of the drawbacks of traditional EDI may be completely eliminated by each solution (Ho, 2008).

Although the removal of regenerant substances is typically viewed as favorable, the compounds do have a minimum of a single benefit. In typical demineralizers, 2%−8% by mass quantities of acid and caustic are added to the IER. The substances clean and rejuvenate the resins simultaneously at these levels. There is less resin cleansing provided by the electrochemical renewal that takes place in a EDI device. For the purpose of to avoid scaling or contamination, adequate preprocessing is thus significantly more crucial when using an EDI device. The need for RO pretreatment for the feed of an EDI system is typically dictated by this, among other things. In general, chemically renewed IERs have less strict feed water specifications than do EDI devices (Wood et al., 2010). Fig. 9.1 depicts the various limitations of EDI technology.

But EDI still faces some serious issues in modern times. In numerous large-scale uses, EDI is frequently much less common than the traditional IE because, for instance, it is costly to set up and hard to maintain due to the use of several pricey IEM and to the complicated apparatus setup. As soon as concentration polarity happens, calcium, magnesium, and heavy metal ions are chemically naturally prone to form precipitates on the outermost layer of the membrane, making the particular IEM used susceptible to mineral scale fouling occurs (Hu et al., 2016). This results in a reduction in deionization performance and a rise in the electrical voltage needed. From a financial and technological perspective, it is obviously very desirable to entirely eliminate these types of membranes. Currently, adopting alternatives such as electrostatically shielded (ES) porous substances is the primary method for addressing issues related to membranes (Shen et al., 2014). When using EDI to purify water that contains hardness or ions of heavy metals, the possibility of scaling or accumulation of metal hydroxides ought to be cautiously avoided (Sun et al., 2016).

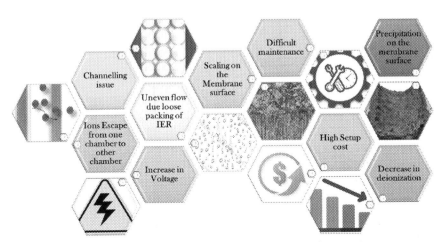

Figure 9.1 Various limitations of EDI technology. *EDI*, Electrodeionization.

9.3 Electrostatic shielding

As an alternative to using permselective IEM, the most recent EDI cell uses electrochemical Faraday cages made of permeable electrically and ionically conductive media or ES CC. They serve as "electrostatic ion pumps" or "ion traps" and electrostatically hold the incoming ions since the chambers locally eliminate the applied electric field, which results in CC. The media with pores are stable both chemically and thermally. Such CC may be substituted for IEM in EDI cells, which are able to be utilized to desalinate brackish or ocean water, renew IER, and create DI water. Without harming the DI procedure, the EDI cells can operate via polarity reversal (PR). The cells are unaffected by existing membrane-associated constraints, such as concentration polarization or scaling, and exhibit increased current effectiveness due to the electronically and ionically active media consisting the ES CC are not permselective and coions are not repelled instead might be took over by the moving counterions (Dermentzis et al., 2012).

Bipolar intermediary electrically and ionically conductive powdered graphite bed electrodes can act as ES zones-ICS and ion CC in MF-EDI uses by eliminating the electrical field that is applied locally within their mass. Thin ES zones − ICS can be used to increase current density while preventing electron transfer redox reactions (Dermentzis et al., 2009b). Table 9.1 gives the application of ES-EDI along with its operating condition and efficiency. Regarding concentration polarization and fouling of membranes, the suggested technique appears attractive and superior to the currently used traditional membrane procedures. It requires more research to be turned into a cutting-edge and economical electrochemical EDI technique for heavy metal elimination (Dermentzis et al., 2011).

Compared to the current traditional EDI technology, this solution appears to be beneficial and attractive in many ways. It requires more study and development to be evolved into a cutting-edge and economical electrochemical technique for desalinating and DI water. To considerably boost the current density and desalination rate while likewise lowering power utilization, thinner porous composite ceramic or polymeric intermediate electrode sheets ought to be used instead of the thick, energy-intensive, self-made powdered graphite beds employed in ES ion CC and ICS. In order to reduce the percentage of the voltage reduction on the two end electrodes, multicells with alternate thin DLC and CC should

Table 9.1 Application of electrostatically shielded electrodeionization (ES-EDI) along with its operating condition and its efficiency.

Application	Specification	ICS material	Operating condition	Efficiency (%)	References
Ammonia removal	Seven compartment EDI Length—15 cm Breadth—12 cm	Electrode graphite powder	Initial concentration—300 ppm Flow rate—5.51×10^{-4} dm^3/s Current density—50 A/m^2 Time—28 min	96.33	Dermentzis et al. (2009b)
Cadmium removal	Seven compartment EDI Length—5 cm Breadth—5 cm	Graphite powder	Initial concentration—50 ppm Flow rate—3.27×10^{-4} L/s Current density—2 mA/cm^2 Time—10 h	99.8	Dermentzis et al. (2011)
Cadmium removal	Five compartment EDI Length—15 cm Breadth—12 cm	Graphite powder	Initial concentration—112.4 ppm Flow rate—2.3×10^{-4} L/s Current density—10.2 A/m^2 Current—0.1 A	99.64	Dermentzis and Ouzounis (2008)

(*Continued*)

Table 9.1 (Continued)

Application	Specification	ICS material	Operating condition	Efficiency (%)	References
Copper removal	Five compartment EDI Length—15 cm Breadth—12 cm	Graphite powder	Initial concentration—63 ppm Flow rate—2.3×10^{-4} L/s Current density—10.2 A/m^2 Current—0.1 A	99.68	Dermentzis and Ouzounis (2008)
Nickel removal	Nine compartment EDI Length—8 cm Breadth—7 cm	Graphite powder	Initial concentration—65.4 ppm Flow rate—2.04×10^{-4} dm^3/s Current density—10.2 A/m^2 Current—0.1 A	99.77	Dermentzis (2010)
Nickel removal	Five compartment EDI Length—15 cm Breadth—12 cm	Graphite powder	Initial concentration—58.7 ppm Current density—0.9 mA/cm^2 Time—20 min Voltage—2 V	73.59	Dermentzis (2008)

Nitrate removal	Seven compartment EDI Length—15 cm Breadth—12 cm	Electrode graphite powder	Initial concentration— 1010 ppm Flow rate— 5.51×10^{-4} dm^3/s Current density— 50 A/m^2 Time—28 min	96.19	Dermentzis et al. (2009a)
Nitrate Removal	Five compartment EDI Length—15 cm Breadth—12 cm	Graphite powder	Initial concentration— 248 ppm Current density— 0.9 mA/cm^2 Voltage—2 V	99.35	Dermentzis (2008)
Potassium removal	Five compartment EDI Length—15 cm Breadth—12 cm	Graphite powder	Initial concentration— 156 ppm Current density— 0.9 mA/cm^2 Voltage—2 V	99.10	Dermentzis (2008)
Sodium chloride removal	Five compartment EDI Length—15 cm Breadth—12 cm	Graphite powder	Initial concentration— 2925 ppm Flow rate— 4×10^{-5} L/s Current density— 10.2 A/m^2 Time—28 min	89.65	Dermentzis and Ouzounis (2008)

(Continued)

Table 9.1 (Continued)

Application	Specification	ICS material	Operating condition	Efficiency (%)	References
Zinc removal	Five compartment EDI Length—15 cm Breadth—12 cm	Graphite powder	Initial concentration— 100 ppm Flow rate— 2.04×10^{-4} dm^3/s Current density— 30 A/m^2 Time—35 min	83	Dermentzis and Ouzounis (2008)
Cobalt removal	Nine compartment EDI Length—8 cm Breadth—7 cm	Graphite powder	Initial concentration— 300 ppm Flow rate— 1.54×10^{-4} dm^3/s Current density— 20 A/m^2	99.7	Dermentzis et al. (2010)

ICS, Ionic current sinks.

also be designed. This will ensure that the electrical energy from the supplied voltage to the cells is used efficiently and nearly solely for the electromigration of ions in the DLC (Dermentzis and Ouzounis, 2008).

9.4 Electrodeionization reversal

Electrode ionization with periodic polarity shifts is referred to as EDIR. The most cost-effective method of producing water for drinking from brackish water appears to be EDIR. At present, EDI is frequently used in the manufacture of highly pure water and the treatment of effluent containing heavy metals. Whenever EDI is used for treating water having hardness or heavy metal ions, the possible danger of scaling or accumulation of heavy metal hydroxides ought to be carefully avoided during the crucial phase of water breakdown in the EDI method. Periodic PR modified the DLC and CC's changing pH characteristics in a way that allowed any scaling or clogging that had developed during the cycle before to break down in the following cycle's slightly acidic circumstances (Sun et al., 2016).

Nickel is extracted and recovered by EDIR or EDI employing PR, from synthetic plating water used for washing. According to the experimental findings, PR should not last more than 4 hours in order to avoid the precipitation of metal hydroxide. Furthermore, the EDIR method with a PR time of 4 hours and sequential flow mode change demonstrated excellent separation efficiency, recuperating these ions into a CC with a high concentration of 3961 ppm while concurrently eliminating 97.0% of nickel from the feed solution. As of now, the efficacy was 32.6%, and treating $1 m^3$ of water used 1.02 kWh of energy. Therefore, EDIR offers a lot of ability for recovering and reusing ions of heavy metals from plating water used to rinse (Lu et al., 2014). Table 9.2 gives the application of EDIR along with its operating condition and efficiency.

As one of the possible strategies to avoid or reduce fouling of membranes and scaling, EDIR utilizing regular PR offers benefits. In this study, the effectiveness of the EDIR method was examined in order to eliminate materials that include divalent cations that cause hardness and to demonstrate its viability as a water-softening method. The influence of operating conditions on process efficacy in the three-phase hydraulic

Table 9.2 Application of electrodeionization reversal (EDIR) along with its operating condition and its efficiency.

Application	Specification	Polarity reversal time	Operating condition	Efficiency (%)	References
Calcium removal	Three stage EDIR Length—12.5 cm Width—8 cm	40 min	Initial concentration—250 ppm Flow rate—20 mL/min Voltage—20 V pH—5.3	99.2	Lee et al. (2012)
Calcium removal	Four compartment EDIR Length—45 cm Width—3.5 cm	60 min	Initial concentration—160 ppm Flow rate—30 mL/min Current—0.85 A pH—5.4	94.56	Yang et al. (2016)
Desalination	Ten compartment EDIR Thickness—0.3 cm	3 h	Initial TDS concentration— 2000 ppm Flow rate—6 L/h Current—1.02 A Voltage—11 V Power consumption— 1.04 kWh/m^3	90.7	Sun et al. (2016)
Desalination	Ten compartment EDIR Thickness—0.3 cm	3 h	Initial TDS concentration— 3000 ppm Flow rate—6 L/h Current—1.54 A Voltage—13 V Power consumption— 1.95 kWh/m^3	90.6	Sun et al. (2016)

Desalination	Ten compartment EDIR Thickness—0.3 cm	3 h	Initial TDS concentration—4000 ppm Flow rate—6 L/h Current—2.18 A Voltage—19 V Power consumption—3.71 kWh/m^3	90.9	Sun et al. (2016)
Magnesium removal	Three stage EDIR Length—12.5 cm Width—8 cm	40 min	Initial concentration—250 ppm Flow rate—20 mL/min Voltage—20 V pH—5.3	99.8	Lee et al. (2012)
Magnesium removal	Four compartment EDIR Length—45 cm Width—3.5 cm	60 min	Initial concentration—96 ppm Flow rate—30 mL/min Current—0.85 A pH—5.4	97.14	Yang et al. (2016)
Nickel removal	Ten compartment EDIR Length—22 cm Width—7.2 cm	4 h	Initial concentration—3961 ppm Flow rate—15 L/h Voltage—30 V Power consumption—1.02 kWh/m^3	97	Lu et al. (2014)
Potassium removal	Four compartment EDIR Length—45 cm Width—3.5 cm	60 min	Initial concentration—156 ppm Flow rate—30 mL/min Current—0.85 A pH—5.4	95.35	Yang et al. (2016)
Sodium removal	Four compartment EDIR Length—45 cm Width—3.5 cm	60 min	Initial concentration—92 ppm Flow rate—30 mL/min Current—0.85 A pH—5.4	90.76	Yang et al. (2016)

EDIR system, such as flow velocity, hardness level, and PR duration. At a flow velocity of 200 mL/min and a calcium carbonate concentration of 250 ppm, a low impedance and excellent elimination effectiveness were seen. Additionally, the findings of the EDIR with increased PR time revealed that the entire range of PR duration was less than 40 minutes without any appreciable pH change. In comparison with the heterogeneous membranes, the homogeneous membrane EDIR operation demonstrated a slightly greater effectiveness for removal. It was discovered that the scale mostly developed on the outermost layer of the CEM as the carbonate form of calcium and magnesium (Lee et al., 2012).

9.5 Membrane-free electrodeionization

Membrane-free EDI (MFEDI) is a cleanup and renewal process that depends on the production of hydroxide ions and hydrogen ions through electrolysis of water and water breakdown. IER may be electrically renewed on-site using MFEDI technique (Shen et al., 2015). It was run in a batch counter-current configuration with small downflow renewal steps and extended upflow maintenance steps alternated. MFEDI offers a simpler setup and does not necessitate the usage of a membrane in comparison with classic EDI (Su et al., 2013; Shen et al., 2019). Table 9.3 gives the application of MF-EDI along with its operating condition and its efficiency. The MFEDI procedure has several benefits. First off, since IEM does not exist in the MFEDI system, issues related to IEM may be eliminated (Jin et al., 2018). Furthermore, MFEDI integrates the benefits of MBIE and EDI. On the other hand, the electrical conductivity of the MFEDI waste water is very close to water that is pure. On the flip side, the IER in the MFEDI column can be electronically renewed in situ, and the resulting concentrates produced when renewal is recyclable to RO. In addition, MFEDI's structure and operation are less complicated than those of MBIE and EDI (Shen et al., 2014).

A minor portion of salt ions that might move to the top new resins under a DC current during the electrical regeneration of the MFEDI eventually led to the product water's purity declining (Shen et al., 2022). During renewal, the electrode polarity has to be switched in order to stop the ions from migrating back. The PR of the electrode polarity significantly decreased renewal effectiveness, complicated renewal, and limited

Table 9.3 Application of membrane-free-electrodeionization along with its operating condition and its efficiency.

Application	Specification	Electrode	Operating condition	Efficiency (%)	References
High purity water production	Double layer resins Resin packing height—500 mm Diameter—30 mm	Reticular electrode	Conductivity—20 μS/cm Flow rate—15 m/h Voltage—730 V pH—7.9 Power consumption—0.68 kWh/m^3 Time—60 h	89	Su et al. (2013)
High purity water production	Double layer resins 1st layer height—70 cm 2nd layer height—30 cm	Reticular electrode	Conductivity—50 μS/cm Flow rate—15 m/h Voltage—1400 V pH—6.9 Power consumption—1.5 kWh/m^3 Time—60 h Current density—280 A/m^2	86	Shen et al. (2014)

(Continued)

Table 9.3 (Continued)

Application	Specification	Electrode	Operating condition	Efficiency (%)	References
High purity water production	Three-layer resins Resin packing height—30 cm Diameter—3 cm	Reticular Ti/ $RuO_2-Sb_2O_5-SnO_2$ electrode	Conductivity— 10 µS/cm Flow rate—15 m/h Voltage—445 V pH—3.4 Power consumption— 0.24 kWh/m^3 Time—72 h	96	Hu et al. (2016)
Nickel removal	Single-layer resins Resin packing height—50 cm Diameter—3 cm	Titanium mesh electrodes	Initial concentration— 10 ppm Flow rate—7.1 L/h Voltage—445 V pH—7.3 Time—72 h Current density— 300 A/m^2	89.5	Shen and Chen (2019)
High purity water production	Single-layer resins Resin packing height—30 cm Diameter—3 cm	Reticular Ti/Pt electrodes	Conductivity— 10 µS/cm Flow rate—15 m/h Voltage—407 V pH—6.5 Power consumption— 0.35 kWh/m^3 Time—48 h	93.1	Hu et al. (2015)

High purity water production	Single-layer resins Resin packing height—40 cm Diameter—3 cm	Reticular Ti/Pt electrodes	Conductivity—5 μS/cm Flow rate—20 m/h Voltage—400 V pH—7 Power consumption— 0.84 kWh/m^3 Time—80 h Current density— 50 A/m^2	74.6	Zhang et al. (2021)
High purity water production	Double layer resins Resin packing height—200 mm Diameter—30 mm	Reticular Ti/RuO$_2$– Sb$_2$O$_5$–SnO$_2$	Conductivity— 20 μS/cm Flow rate—14 L/h Voltage—885 V pH—7.2 Power consumption— 0.71 kWh/m^3 Time—20 min Current density— 150 A/m^2	90	Su et al. (2014)

(Continued)

Table 9.3 (Continued)

Application	Specification	Electrode	Operating condition	Efficiency (%)	References
Nickel removal	Single layer resins height—500 mm Diameter—30 mm	Reticular Ti/Pt electrodes	Conductivity—42.5 µS/cm Initial concentration—10 ppm Flow rate—10 m/h Voltage—1235 V pH—7.4 Power consumption—5.6 kWh/m^3 Time—80 h Current density—300 A/m^2	88	Shen et al. (2015)
Nickel removal	Double layer resins height—500 mm Diameter—30 mm	Reticular Ti/Pt electrodes	Initial concentration—10 ppm Flow rate—10.6 L/h Voltage—1215 V pH—7.5 Power Consumption—4 kWh/m^3 Time—20 min Current density—300 A/m^2	87.5	Shen et al. (2019)

electrode lifetime. Although a way of blocking the transport route of cation ions by utilizing several thin AER layers was established, this technology was challenging to use in practical situations due to the poor efficiency and the complex functioning (Hu et al., 2015). Consequently, a straightforward and efficient enhancement strategy is provided for preventing electrode PR and improving renewal performance (Su et al., 2014). The IER layer design was enhanced, and the proportions of CER and AER were adjusted over the whole IER layer, to stop salt ions from migrating back. The MFEDI was mostly loaded with combined strong-acid and strong-base resins, which offer good desalination efficiency, while purifying the somewhat acidic RO effluent (Hu et al., 2016).

The addition of phosphonic acid resin considerably increased the ability of MFEDI to remove nickel ions. The effluent's conductivity was less than 0.3 S/cm, which indicated significant purifying effectiveness. The average quantity of concentrated nickel was over 95 mg/L, which is indicative of effective electrical renewal. Using a multilayer composite IER bed, the concentrate's pH was effectively brought down to neutral, allowing for recycling. In addition, no nickel hydroxide precipitated was produced during the tests. Additionally, it was discovered that MFEDI with phosphonic acid resin used less than 73% of the energy than MFEDI. The electrical renewal method was also put forth. The splitting of water was responsible for more than 72% of the renewal of IER, whereas electrolysis of water was only responsible for less than 28% of the renewal of IER. After 10 runs of procedure, it was clear that the system's efficacy for the treatment and recuperation of effluent comprising nickel was consistent and reproducible (Shen and Chen, 2019).

Conventional IERs that cannot concurrently accomplish efficient adsorption, simple electrical renewal, and strong conductance are the functional materials employed in MFEDI. For example, strong acid IER, a CER frequently utilized in MFEDI, is not readily electrically renewed; weak acid resin, on the other hand, may be quickly renewed but has poor conductance and adsorption qualities (Zhang et al., 2022). There are not any IERs on the market right now that can completely satisfy the demands for conductivity, adsorption, and renewal standards. The addition of anionic functional groups caused a new amphoteric resin to behave as a medium-strength acid resin, even though it had a lot more cation functional groups than anions functional groups. The synthesized amphoteric resin had superior purifying capabilities compared to weak acid resin and was also easier to electrical regeneration (Zhang et al., 2021).

9.6 Resin wafer electrodeionization

Ultrapure water may be created using EDI in the medical and semiconductors sectors. Due to the scattered IER beads used, traditional EDI has inconsistent efficiency and challenging on-site servicing. The speed of ion elimination may be reduced by channel development brought on by loosened IER beads. The stack needs to be treated carefully in a controlled atmosphere when using scattered IER (Lopez and Hestekin, 2015; Du et al., 2012). The free IER beads were swapped out by resin wafers made of the original scattered IEX resins that had been immobilized and molded into a transparent, solid base in order to advance EDI technique. Ionic movement is enhanced by RW-EDI, and local pH control is also possible (Pan et al., 2017; Ho et al., 2010). In terms of production and energy consumption, saltwater desalination with RW-EDI ought to be comparable with commercialized RO (Hestekin et al., 2020). In the instance of cooling water restoration at thermoelectric facilities in the United States, adopting RW-EDI for cooling water desalination rather than commercialized RO could conserve a significant amount of power (Tseng et al., 2022; Jordan et al., 2021). More than 99% of the salt can be removed within 120 minutes, and the elimination of salt proportion rises with time. The rise in input flow velocity from 410 to 840 mL/min had no effect on the elimination of salt because batch-mode operation was used. On the other hand, the voltage applied has a less of an impact on energy use and performance than input rate (Pan et al., 2018). Fig. 9.2 depicts the schematic representation of RW-EDI.

RW-EDI was utilized for a feed salt level of 3000 ppm at various voltages being applied and input flow rates. The findings suggested that 94% saline elimination efficacy might be attained. A first-order kinetic equation was used to study the NaCl elimination dynamics from the saltwater. At a cell potential of 2.53 V, it was discovered that the highest kinetic rate constant was 0.091 per min. The ideal working parameters ought to be adjusted at an applied voltage of 2.28 V per cell pair and a flow rate for the feed of 810 mL per minute, which results in a yield of 55.5 L per hour per square meter at a consumption of energy of $0.66 \, kWh/m^3$. RW-EDI can provide a saltwater desalination efficiency that is considerably more environmentally friendly in comparison with other water desalination procedures (Zheng et al., 2018).

Advances future scope in electrodeionization

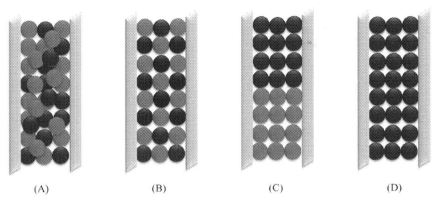

Figure 9.2 Schematic representation of resin wafer electrodeionization (RW-EDI): (A) conventional EDI, (B) mixed bed RW-EDI, (C) layered bed RW-EDI, and (D) anion exchange RW-EDI. *EDI*, Electrodeionization.

CO_2 from fossil fuel-fired power station exhaust gas might be effectively captured using a unique RE-EDI technique. The technique uses consecutive compartments to continuously swing the method's liquid's pH from basic to acidic while utilizing in situ electrical pH control with RW-EDI (Lin et al., 2014). The range of the pH fluctuation was significantly influenced by the current that was used and liquid velocity. The CO_2 may be trapped in the basic compartment owing to the pH fluctuation and then released in the acidic compartment. The technique offers a mechanism for recuperating CO_2 at standard pressure without the need for heating up, vacuuming or use of chemicals. From a fake flue gas, up to 80% carbon dioxide was recovered with greater 98% pure. Hours of sustained performance were seen, indicating the device could deliver desired performance (Datta et al., 2013).

RW-EDI is regarded as an attractive technique for the desalination of seawater that uses less energy. However, the energy needs for the treatment of feeds with high salinities place a restriction on RW-EDI and preclude its widespread use (Jordan et al., 2023). The efficacy of RW-EDI as a consequence is influenced by a number of significant aspects. The rigidity of resin wafers is now ensured by a nonconductive polymeric binder in cutting-edge wafers. Yet, the RW conductance and power consumption of RW-EDI systems might be enhanced by a strong ionic polymers (Palakkal et al., 2020). In order to improve ion conductance during RW-EDI procedure, polymer ionic liquid (PIL)-based ionic polymers were added to the compositions of IER wafers (Fasuyi, 2019). Ionic polymers based on polymer ionic liquids were used in the RW-EDI system to reduce particular consumption of energy by 9.3% as compared to standard RW-EDI (Fasuyi and Lopez, 2023). The findings show that PIL ionic polymers may considerably raise RW-EDI system energy consumption.

9.7 Coupling of electrodeionization with other techniques

For removing boron from solution, a hybrid method made of polymer assisted ultrafiltration and EDI was created (Arar et al., 2013b; Kresnowati et al., 2019). As a preliminary step, polymer assisted ultrafiltration was utilized to lower the amount of boron content and boost the efficiency of the EDI device. A boron elimination rate of 68% by polymer-assisted ultrafiltration was achieved in the preprocessing step by optimizing the copolymer/boron ratio, solution pH, and the interaction duration. It was then moved to the EDI unit with the previously treated mixture. The rate of flow and the voltage that was used both had an impact on the rate of elimination of boron during the EDI process, while the conductance of the solution of sodium sulfate in the electrode chamber had no effect. At a flow rate of 2 L/h and an applied voltage of 30 V, the best rate of elimination was attained. The elimination rate was enhanced to more than 99.9% by extending the working period from 3 to 10 hours and raising the total number of layers in the EDI unit. The findings demonstrate that a potential method to lower the boron content in ultrapure water is the polymer aided ultrafiltration EDI system (Soysüren et al., 2023).

To provide exceptionally pure water, a hybrid RO-CEDI technique was studied. The CEDI unit trials were conducted in a cell-pair stacking with three separate compartments, while the RO system, with an effective membrane area of 1.1 m^2, was run using municipal water. The RO extract has a conductance of less than 20 μs/cm, whereas the feed is ground water with a conductance of roughly 400 μs/cm. With a resistance of 10−16.7 MΩ cm, the generated water satisfies the standards for quality as the water used for makeup in a nuclear power plant (Song et al., 2005). Fig. 9.3 shows the schematic representation of hybrid RO/EDI system. For the removal of minerals of RO filtrate from geothermal water, a mixed bed EDI system was used. The outcomes showed that

Advances future scope in electrodeionization

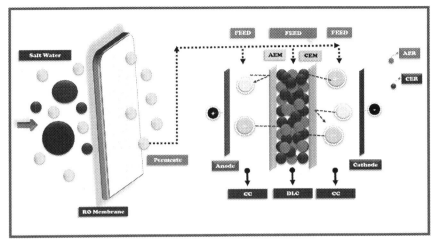

Figure 9.3 Schematic representation of hybrid RO/EDI system. *EDI*, Electrodeionization; *RO*, reverse osmosis.

applying excessive voltage to the EDI system enhanced the conductive ability of the product water. The electric conductivity of the end product water reduced in EDI operations when thick membranes were utilized. The product water's conductance increased as a consequence of employing thin membranes. It was decided that if additional demineralization of geothermal water is required after the RO process in order to employ it in the system for heating, the EDI approach might be taken into consideration as a feasible method (Arar et al., 2013a, 2014b).

Using an artificial, diluted sugar solution as an example for sugar-containing effluent, RO-EDI is utilized to simultaneously concentrate sugar and remove salts. A combined thin film RO membrane is used in the RO system. One anode chamber, one cathode chamber, two DLC, and two CC are present in the EDI stack. The EDI stack's IEMs are made of readily accessible CEM and AEM. Both DLC and CC are packed with a mixture of IER. Research results demonstrate that there's not any sugar waste in the EDI stack and that the measured sugar reject of the RO membrane is greater than 99.9%. This shows that virtually all of the sugar can be recovered using the hybrid approach. Additionally, the highly concentrated sugar solution's salt level is significantly reduced. However, the EDI-RO design appears to be superior than the RO-EDI arrangement in terms of permeate flow and permeate purity (Widiasa and Wenten, 2007).

9.8 Artificial intelligence in electrodeionization

Artificial intelligence (AI) has gotten a lot of interest from government agencies, businesses, and research. The study of AI focuses on how to program computers to carry out intellectual activities that, in past times, could only be completed by people (Xu et al., 2021; Parunak, 1996). AI has advanced quickly in recent years, changing the way people live. Increasing the nation's competitiveness and safeguarding protection, the creation of AI has grown into a crucial component of international development strategies (Zhang and Lu, 2021). Despite significant advancements in the field of machine learning (Chassignol et al., 2018), ANN still fall short of human brains in their capacity to generalize to novel circumstances. Different generalizations for identical training information are induced by a variety of distinguishing characteristics of an educational system, including its network structure as well as the learning rule. The combined result of these two factors, known as "inductive bias," affects the accuracy with which any learning algorithm—or cerebrum—generalizes. Robust generalization requires strong inductive biases. Artificial networks frequently grab onto patterns that only provide information about the statistics of the training data and are not always generalizable to other contexts. On the other hand, brains constantly generalize across very significant changes in the sensory input (Sinz et al., 2019).

The modeling of RW-EDI using a multiion solution with a substantial amount has been completed. For the modeling of RW-EDI with high quantities of multiion solutions, data science and machine learning approaches are used. The amounts of calcium, sodium, and magnesium ions in the solution were estimated with multioutput a regression analysis and artificial neural network multiple-layer perceptron, and the effects of various RW were verified using both multioutput and a single-output regression analysis, leave-one-out cross validation, and neural networks multilayer perceptron (Zhou et al., 2022; Erol, 2021). ANN was used to provide optimal experimental conditions in order to achieve the maximum Sr and Cs removal percentage. For this purpose, experimental data were used to train the two separate ANNs. The control factors (feed concentration, flow rate, and voltage) were utilized as the inputs of ANNs and Sr and Cs removal percentages were used as the outputs of the two ANNs. The consequences demonstrated that there was a good agreement between the experimental and predicted results (Zahakifar et al., 2017).

9.9 Conclusion

The EDI method is examined in this chapter, along with its benefits and drawbacks. Additionally, it looked at the advantages of the most current developments in EDI such as EDIR, MF-EDI, RW-EDI, ES-EDI, and hybrid method along with EDI. The use of EDI for elimination, segregation, and concentrating has so far been demonstrated to be good and ecologically friendly. In order to advance and develop this field of study, new materials have been continually created. If successful, this kind of technology might have significant worldwide financial and ecological advantages. The development of new electrodes coatings with greater activity and larger surface areas, as well as ES, RW, and nanotechnology, will lead to the possibility for dramatically raising the efficacy of EDI devices. Because they use less energy, EDI systems are projected to continue to advance and show more efficiency while running more affordably, making them desirable for industrial scale-up across a variety of applications globally.

References

Alvarado, L., Chen, A., 2014. Electrodeionization: principles, strategies and applications. Electrochimica Acta 132, 583–597.

Arar, Ö., Yüksel, Ü., Kabay, N., Yüksel, M., 2013a. Demineralization of geothermal water reverse osmosis (RO) permeate by electrodeionization (EDI) with layered bed configuration. Desalination 317, 48–54.

Arar, Ö., Yüksel, Ü., Kabay, N., Yüksel, M., 2013b. Application of electrodeionization (EDI) for removal of boron and silica from reverse osmosis (RO) permeate of geothermal water. Desalination 310, 25–33.

Arar, Ö., Yüksel, Ü., Kabay, N., Yüksel, M., 2014a. Various applications of electrodeionization (EDI) method for water treatment—a short review. Desalination 342, 16–22.

Arar, Ö., Yüksel, Ü., Kabay, N., Yüksel, M., 2014b. Demineralization of geothermal water reverse osmosis (RO) permeate by electrodeionization (EDI) with mixed bed configuration. Desalination 342, 23–28.

Boontawan, P., Kanchanathawee, S., Boontawan, A., 2011. Extractive fermentation of l-(+)-lactic acid by Pediococcus pentosaceus using electrodeionization (EDI) technique. Biochemical Engineering Journal 54 (3), 192–199.

Chassignol, M., Khoroshavin, A., Klimova, A., Bilyatdinova, A., 2018. Artificial Intelligence trends in education: a narrative overview. Procedia Computer Science 136, 16–24.

Datta, S., Henry, M.P., Lin, Y.J., Fracaro, A.T., Millard, C.S., Snyder, S.W., et al., 2013. Electrochemical CO_2 capture using resin-wafer electrodeionization. Industrial & Engineering Chemistry Research 52 (43), 15177–15186.

Dermentzis, K., Ouzounis, K., 2008. Continuous capacitive deionization–electrodialysis reversal through electrostatic shielding for desalination and deionization of water. Electrochimica Acta 53 (24), 7123–7130.

Dermentzis, K., 2008. Continuous electrodeionization through electrostatic shielding. Electrochimica Acta 53 (6), 2953–2962.

Dermentzis, K., 2010. Removal of nickel from electroplating rinse waters using electrostatic shielding electrodialysis/electrodeionization. Journal of Hazardous Materials 173 (1–3), 647–652.

Dermentzis, K., Papadopoulou, D., Christoforidis, A., Dermentzi, A., 2009a. A new process for desalination and electrodeionization of water by means of electrostatic shielding zones-ionic current sinks. Journal of Engineering Science & Technology Review 2 (1).

Dermentzis, K., Davidis, A., Papadopoulou, D., Christoforidis, A., Ouzounis, K., 2009b. Copper removal from industrial wastewaters by means of electrostatic shielding driven electrodeionization. Journal of Engineering Science & Technology Review 2 (1).

Dermentzis, K.I., Davidis, A.E., Dermentzi, A.S., Chatzichristou, C.D., 2010. An electrostatic shielding-based coupled electrodialysis/electrodeionization process for removal of cobalt ions from aqueous solutions. Water Science and Technology 62 (8).

Dermentzis, K., Christoforidis, A., Papadopoulou, D., Davidis, A., 2011. Ion and ionic current sinks for electrodeionization of simulated cadmium plating rinse waters. Environmental Progress & Sustainable Energy 30 (1), 37–43.

Dermentzis, K., Davidis, A., Chatzichristou, C., Dermentzi, A., 2012. Ammonia removal from fertilizer plant effluents by a coupled electrostatic shielding based electrodialysis/electrodeionization process. Global NEST Journal 14, 468–476.

Du, J., Lorenz, N., Beitle, R.R., Hestekin, J.A., 2012. Application of wafer-enhanced electrodeionization in a continuous fermentation process to produce butyric acid with Clostridium tyrobutyricum. Separation Science and Technology 47 (1), 43–51.

Erol, H.U., 2021. Investigative Study of Microalgal and Electrochemical Wastewater Treatment Systems and Modeling of the Wafer-Enhanced Electrodeionization Using Supervised Learning. University of Arkansas.

Ervan, Y., Wenten, I.G., 2002. Study on the influence of applied voltage and feed concentration on the performance of electrodeionization. Songklanakarin Journal of Science and Technology 24, 955–963.

Fasuyi, A., 2019. Investigative Study of Poly (Ionic) Liquid Incorporation in Resin Wafer Electrodeionization for Improved Specific Energy Consumption.

Fasuyi, A., Lopez, A.M., 2023. Influence of poly (ionic) liquid incorporation within resin wafer electrodeionization for reduced energy consumption in brackish water desalination. Chemical Engineering Journal 454, 140209.

Gifford, J.D., Atnoor, D.E.V.E.N., 2000. October. An innovative approach to continuous electrodeionization module and system design for power applications. International Water Conference. Oetober, Pittsburg, PA, pp. 22–26.

Hestekin, C.N., Hestekin, J.A., Paracha, S., Morrison, G., Pakkaner, E., Moore, J., et al., 2020. Simulating nephron ion transport function using activated wafer electrodeionization. Communications Materials 1 (1), 20.

Ho, T., 2008. Ion Selective Removal Using Wafer Enhanced Electrodeionization. University of Arkansas.

Ho, T., Kurup, A., Davis, T., Hestekin, J., 2010. Wafer chemistry and properties for ion removal by wafer enhanced electrodeionization. Separation Science and Technology 45 (4), 433–446.

Hu, J., Fang, Z., Jiang, X., Li, T., Chen, X., 2015. Membrane-free electrodeionization using strong-type resins for high purity water production. Separation and Purification Technology 144, 90–96.

Hu, J., Chen, Y., Zhu, L., Qian, Z., Chen, X., 2016. Production of high purity water using membrane-free electrodeionization with improved resin layer structure. Separation and Purification Technology 164, 89–96.

Jin, X., Zhou, C., He, Z., Lou, X., Yu, C., Chen, X., 2018. Membrane-free electrodeionization for high-velocity production of high-purity water. Desalination and Water Treatment 126, 24–31.

Jordan, M.L., Palakkal, V.M., Lin, Y., Arges, C.G., Valentino, L., 2021. Electrodeionization of organic acids using porous bipolar resin wafers, Electrochemical Society Meeting Abstracts, 239. The Electrochemical Society, Inc, p. 970, No. 27.

Jordan, M.L., Kokoszka, G., Gallage Dona, H.K., Senadheera, D.I., Kumar, R., Lin, Y.J., et al., 2023. Integrated Ion-Exchange Membrane Resin Wafer Assemblies for Aromatic Organic Acid Separations Using Electrodeionization. ACS Sustainable Chemistry & Engineering.

Khoiruddin, K., Hakim, A.N., Wenten, I.G., 2014. Advances in electrodeionization technology for ionic separation—a review. Membrane and Water Treatment 5 (2), 87–108.

Kresnowati, M.T.A.P., Regina, D., Bella, C., Wardani, A.K., Wenten, I.G., 2019. Combined ultrafiltration and electrodeionization techniques for microbial xylitol purification. Food and Bioproducts Processing 114, 245–252.

Kumar, P.S., Varsha, M., Rathi, B.S., Rangasamy, G., 2022. Electrodeionization: Fundamentals, methods and applications. Environmental Research 114756.

Lee, H.J., Hong, M.K., Moon, S.H., 2012. A feasibility study on water softening by electrodeionization with the periodic polarity change. Desalination 284, 221–227.

Lee, H.J., Song, J.H., Moon, S.H., 2013. Comparison of electrodialysis reversal (EDR) and electrodeionization reversal (EDIR) for water softening. Desalination 314, 43–49.

Lin, Y.J., Snyder, S.W., Trachtenberg, M.C., Cowan, R.M., Datta, S., 2014. Carbon Dioxide Capture using Resin-Wafer Electrodeionization. Argonne National Lab. (ANL), Argonne, IL, No. 8,864,963.

Lopez, A.M., Hestekin, J.A., 2015. Improved organic acid purification through wafer enhanced electrodeionization utilizing ionic liquids. Journal of Membrane Science 493, 200–205.

Lu, H., Wang, Y., Wang, J., 2014. Removal and recovery of Ni^{2+} from electroplating rinse water using electrodeionization reversal. Desalination 348, 74–81.

Palakkal, V.M., Valentino, L., Lei, Q., Kole, S., Lin, Y.J., Arges, C.G., 2020. Advancing electrodeionization with conductive ionomer binders that immobilize ion-exchange resin particles into porous wafer substrates. npj Clean Water 3 (1), 5.

Pan, S.Y., Snyder, S.W., Ma, H.W., Lin, Y.J., Chiang, P.C., 2017. Development of a resin wafer electrodeionization process for impaired water desalination with high energy efficiency and productivity. ACS Sustainable Chemistry & Engineering 5 (4), 2942–2948.

Pan, S.Y., Snyder, S.W., Ma, H.W., Lin, Y.J., Chiang, P.C., 2018. Energy-efficient resin wafer electrodeionization for impaired water reclamation. Journal of Cleaner Production 174, 1464–1474.

Parunak, H.V.D., 1996. Applications of distributed artificial intelligence in industry. Foundations of Distributed Artificial Intelligence 2 (1), 18.

Rathi, B.S., Kumar, P.S., 2020. Electrodeionization theory, mechanism and environmental applications. A review. Environmental Chemistry Letters 18, 1209–1227.

Rathi, B.S., Kumar, P.S., Parthiban, R., 2022. A review on recent advances in electrodeionization for various environmental applications. Chemosphere 289, 133223.

Shen, X., Chen, X., 2019. Membrane-free electrodeionization using phosphonic acid resin for nickel containing wastewater purification. Separation and Purification Technology 223, 88–95.

Shen, X., Li, T., Jiang, X., Chen, X., 2014. Desalination of water with high conductivity using membrane-free electrodeionization. Separation and Purification Technology 128, 39–44.

Shen, X., Fang, Z., Hu, J., Chen, X., 2015. Membrane-free electrodeionization for purification of wastewater containing low concentration of nickel ions. Chemical Engineering Journal 280, 711–719.

Shen, X., Yu, J., Chen, Y., Peng, Z., Li, H., Kuang, X., et al., 2019. Influence of some parameters on membrane-free electrodeionization for the purification of wastewater containing nickel ions. International Journal of Electrochemical Science 14, 11237–11252.

Shen, X., Liu, Q., Li, H., Kuang, X., 2022. Membrane-free electrodeionization using graphene composite electrode to purify copper-containing wastewater. Water Science & Technology 86 (7), 1733–1744.

Sinz, F.H., Pitkow, X., Reimer, J., Bethge, M., Tolias, A.S., 2019. Engineering a less artificial intelligence. Neuron 103 (6), 967–979.

Song, J.H., Yeon, K.H., Cho, J., Moon, S.H., 2005. Effects of the operating parameters on the reverse osmosis-electrodeionization performance in the production of high purity water. Korean Journal of Chemical Engineering 22, 108–114.

Soysüren, G., İpek, İ.Y., Arda, M., Arar, Ö., 2023. Polymer-enhanced ultrafiltration–electrodeionization hybrid system for the removal of boron. In: Environmental Science: Water Research & Technology.

Strathmann, H., 2010. Electrodialysis, a mature technology with a multitude of new applications. Desalination 264 (3), 268–288.

Su, W., Pan, R., Xiao, Y., Chen, X., 2013. Membrane-free electrodeionization for high purity water production. Desalination 329, 86–92.

Su, W., Li, T., Jiang, X., Chen, X., 2014. Membrane-free electrodeionization without electrode polarity reversal for high purity water production. Desalination 345, 50–55.

Sun, X., Lu, H., Wang, J., 2016. Brackish water desalination using electrodeionization reversal. Chemical Engineering and Processing: Process Intensification 104, 262–270.

Tseng, P.C., Lin, Z.Z., Chen, T.L., Lin, Y., Chiang, P.C., 2022. Performance evaluation of resin wafer electrodeionization for cooling tower blowdown water reclamation. Sustainable Environment Research 32 (1), 36.

Ulusoy Erol, H.B., Hestekin, C.N., Hestekin, J.A., 2021. Effects of resin chemistries on the selective removal of industrially relevant metal ions using wafer-enhanced electrodeionization. Membranes 11 (1), 45.

Wang, J., Wang, S., Jin, M., 2000. A study of the electrodeionization process—high-purity water production with a RO/EDI system. Desalination 132 (1–3), 349–352.

Wardani, A.K., Hakim, A.N., Wenten, I.G., 2017. Combined ultrafiltration-electrodeionization technique for production of high purity water. Water Science and Technology 75 (12), 2891–2899.

Wenten, I.G., Arfianto, F., 2013. Bench scale electrodeionization for high pressure boiler feed water. Desalination 314, 109–114.

Widiasa, I.N., Wenten, I.G., 2007. Combination of reverse osmosis and electrodeionization for simultaneous sugar recovery and salts removal from sugary wastewater. Reaktor 11 (2), 91–97.

Widiasa, I.N., Sutrisna, P.D., Wenten, I.G., 2004. Performance of a novel electrodeionization technique during citric acid recovery. Separation and Purification Technology 39 (1–2), 89–97.

Wood, J., Gifford, J., Arba, J., Shaw, M., 2010. Production of ultrapure water by continuous electrodeionization. Desalination 250 (3), 973–976.

Xu, Y., Liu, X., Cao, X., Huang, C., Liu, E., Qian, S., et al., 2021. Artificial intelligence: a powerful paradigm for scientific research. The Innovation 2 (4), 100179.

Yang, G., Zhang, Y., Guan, S., 2016. Study on the desalination of high hardness water by electrodeionization reversal. Desalination and Water Treatment 57 (18), 8127–8138.

Zahakifar, F., Keshtkar, A., Nazemi, E., Zaheri, A., 2017. Optimization of operational conditions in continuous electrodeionization method for maximizing Strontium and Cesium removal from aqueous solutions using artificial neural network. Radiochimica Acta 105 (7), 583–591.

Zeng, G., Ye, J., Yan, M., 2016. Application of electrodeionization process for bioproduct recovery and CO_2 capture and storage. Current Organic Chemistry 20 (26), 2790–2798.

Zhang, C., Lu, Y., 2021. Study on artificial intelligence: the state of the art and future prospects. Journal of Industrial Information Integration 23, 100224.

Zhang, X., Deng, S., Jin, H., Yu, Y., Liao, S., Chen, X., 2021. Synthesis and characterization of an amphoteric resin for use in membrane-free electrodeionization. Separation and Purification Technology 272, 118857.

Zhang, X., Jin, H., Deng, S., Xie, F., Li, S., Chen, X., 2022. Amphoteric blend ion exchange resin with medium-strength alkalinity for high-purity water production in membrane-free electrodeionization. Desalination 529, 115663.

Zheng, X.Y., Pan, S.Y., Tseng, P.C., Zheng, H.L., Chiang, P.C., 2018. Optimization of resin wafer electrodeionization for brackish water desalination. Separation and Purification Technology 194, 346–354.

Zhou, X., Yan, G., Majdi, H.S., Le, B.N., Khadimallah, M.A., Ali, H.E., et al., 2022. Spotlighting of microbial electrodeionization cells for sustainable wastewater treatment: application of machine learning. Environmental Research 115113.

CHAPTER TEN

Economics and environmental aspects of the electrodeionization technique

10.1 Introduction

Electrodeionization (EDI) is a technique for reducing concentration polarization (CP) in electrodialysis (ED) devices (Saravanan et al., 2023). The benefits of ED and ion–exchange (IE) technologies are merged via a beneficial combination of ED and IE, which may successfully solve various performance problems linked to each approach independently (Alvarado and Chen, 2014). In general, in comparison with other inorganic solid elimination techniques, EDI has a number of benefits, such as a complete recovery of the aimed pollutants, a greater IE membrane (IEM) life than the process of reverse osmosis (RO), a more time IE resin (IER) life in comparison with other operations, and an easier IE method for IER recovery. Compared to other conventional wastewater treatment (WWT) methods, EDI offers a number of benefits, notably easy and reliable operation, complete elimination of the necessity for renewal chemicals, cheap operating and maintenance costs, low energy consumption, nonpollution, safety, and reliability. The process takes up little space and requires few automatic valves or intricate control sequencing that need operator supervision (Wood et al., 2010; Rathi and Kumar, 2020). It also continually produces high purity water (HPW).

The method of design for EDI and traditional ED is quite similar. The primary distinction is that in an EDI stack, IER is placed in the dilute cells (DLCs) and occasionally the concentrate cells (CC). The arrangement of cation exchange resin (CER) and anion exchange resin (AER) in the cell is done using several theories (Rathi et al., 2022). According to Wang et al. (2000), after being adsorbate by the IER, both cations and anions are subsequently transported by an electrical potential (EP) gradient across the appropriate IER in the direction of the neighboring CC confronting the cathode

Electrodeionization: Fundamentals, Methods and Applications.
DOI: https://doi.org/10.1016/B978-0-443-18983-8.00010-7
© 2024 Elsevier Inc. All rights reserved, including those for text and data mining,
AI training, and similar technologies.

and the anode, correspondingly. The stack may be handled cheaply at a fairly high current density as compared to traditional ED because the ion conductance in the IER is many orders of magnitude greater than that in the deionized water (DW) (Bouhidel and Lakehal, 2006). Still, the elimination of weak bases and acids such as boric or silicic acid is fairly poor when a mixed-bed IER (MIER) is used in EDI. In a setup where the CER and AER are situated in distinct chambers with a bipolar membrane (BM), poorly dissociated electrolytes may be eliminated much more effectively (Wenten and Arfianto, 2013; Hakim and Wenten, 2014). To assess water output, usage of energy, and financial viability, EDI must be assessed. According to yearly expenses, which include energy costs, fixed costs, labor costs, IEM replacement costs, repair expenses, reagent costs, and insurance premiums, specific product water costs are calculated (Generous et al., 2021). To comprehend the primary economic factors and obstacles, comprehensive technoeconomic analysis (TEA) and life cycle assessment (LCA) are used (Valentino et al., 2022). The objective of this chapter was to examine into the effects of EDI on safety and health. Furthermore, it covers at the EDI design, along with its technoeconomic assessment (TEA) and life cycle analysis (LCA). A key element in the effective implementation of EDI is friendly to the environment.

10.2 Electrodeionization

Water ionized species are eliminated using EDI, which integrates IER, IEM, and DC. It was created to get around the IE system's shortcomings, which include the necessity for IER renewal and the production of not satisfactory product in ED. In comparison with the traditional IE system, EDI has the benefit of being a continuous process with a consistent level of quality, which is able to create HPW eliminating a requirement for acid or caustic renewal. According to Wardani et al. (2017), the most popular method for generating HPW is EDI, which generally results in rejection of salt rates of above 99.5% with a resistance of $1-18$ M cm and minimal TOC.

In an EDI device, cation exchange membranes (CEMs) and anion exchange membranes (AEMs) alternately sit among the pair of electrodes. DLC, or product chambers, CC, and EC make up the EDI stack's chambers. MIER is used to fill every compartment, which improves the transfer of ionic elements from the entire solution to the IEM when a DC is applied (Alvarado et al., 2013). The IER beads improve mass transfer, enable water splitting, and lessen stack impedance when EDI is employed to produce HPW (Alvarado et al., 2009).

Water is divided into hydrogen and hydroxyl ions by the DC electric field, and this process constantly regenerates the IER. The membranes convey the newly exchanged ions to the CC where they are drained out of the system (Qian et al., 2022). The qualities of the feed have a big impact on the level of quality of the end product water produced by the EDI method. The calcium carbonate content of feed water must typically be below one ppm in order to use the current EDI method (Wen et al., 2005). RO is therefore frequently required to be used for the EDI's preliminary processing (Lee et al., 2007). Ultrafiltration (UF) is typically required as a pretreatment for RO in order to remove particles that could potentially clog or harm the RO membrane. Additionally, RO demands a very high working pressure, which increases the degree of complexity of the piping and instrumentation and increases the consumption of energy (Kumar et al., 2022; Fu et al., 2009).

However, nowadays, EDI still suffers from some harsh problems. For example, it is costly in investment and difficult in repairing due to the usage of multiple expensive IEM and to the complex equipment configuration, which often renders EDI much less popular than the conventional IE in many large-scale applications (Verbeek et al., 1998). Also, the selective IEM used is prone to mineral-scale fouling because calcium, magnesium, and heavy metal (HM) ions are chemically inclined to precipitate on the membrane surface once the CP occurs, leading to a decrease in deionization efficiency and an increase in the EP required. From an economic and technological perspective, it is thus very desirable to entirely eliminate these IEM. The major method used to address IEM-related issues at the moment is the use of alternatives such electrostatically shielded EDI (ES-EDI) porous substances. This can partially address the issue, but the EDI that uses the replacements often has low purifying efficiency and a complex design (Su et al., 2013).

10.3 Health aspects in electrodeionization

The related industrialization, urbanization, and chemically improved agriculture that come along with an expanding population all add to the contamination of already overtaxed water supplies (Muoz and Guieysse, 2006). Some of the contaminants that are of special concern to people of all ages are organic compounds, HM, vitamins and minerals, colors, medicine, polyfluoroalkyl, and perfluoroalkyl chemicals, radioactive substances, plastic goods, tiny particles, and microorganisms (Zhang et al., 2020). Numerous individuals throughout the world are finding it more and more difficult to access drinking water, and the problem becomes especially severe in poorer nations (Jeong et al., 2022). Due to this, it is imperative to develop resources and techniques that are affordable, accessible, ecologically friendly, thermally effective, broadly available, and chemically resistant in order to meet the world's growing need for safe water to consume (Wu et al., 2010).

For the elimination of various aquatic contaminants, several water and WWT methods were created in the recent years. Due to their benefits, such as ecological compatibility, adaptability, excellent energy efficiency, flexibility of automation and protection, and cost efficacy, electrochemical techniques (ECTs) have been researched and implemented as different possibilities for the WWT over the past few years (Pulkka et al., 2014). ECT may be employed to eliminate a wide variety of contaminants from WW. According to Suman et al. (2021), ECT for WW is becoming more and more popular due to it has a number of benefits over traditional techniques, including total mineralization, a need for little to no reagents made from chemicals, the prevention of the emergence of new harmful substances, and low cost of energy. The flexibility to utilize the same procedure with many types of WW, automation suitability, and affordability are some further benefits of ECT (Särkkä et al., 2015). The drawbacks include that initial expenditure is costly, and corrosion of the electrodes can occasionally happen. Additionally, when WW has a high chloride content, electrochemical oxidation of pollutants has shown the results with great efficiency (Alkhadra et al., 2022).

For the DS of seawater and the concentration of compounds with added value in WW, EDI is a separation technique that has been extensively utilized (Song et al., 2007). Researchers are concerned about the presence of metals in the water since it has sharply grown in the past few years (Rathi et al., 2021). It is challenging to eliminate HM and other pollutants from water because they may react either when in their purest chemical state or in conjunction with other ions (Bilal et al., 2019). Policies and technological solutions must be created in order to solve this problem without having potentially severe effects on the well-being of humans and the ecosystem.

Ion species can be eliminated by EDI from solutions that have been diluted. Both the ion transport and the electrically conductive properties of the solution are improved by the inclusion of IER in the DLC. Aqueous solutions comprising different metal ions, including Ni, Cs, Cu, B, and Si, as well as Zn, Co, and Sr, have been effectively treated using EDI. Without the use of chemicals, these charged particles elute IER. Since no chemicals are needed in the EDI method to elute the IER, this technology is ecologically safe. In comparison with the traditional IE process, EDI has lower operational costs. Essential benefits of the EDI approach in industrial applications include avoiding the incorporation of chemical substances to the extraction of IER and removing the expenses associated with acquiring, transporting, and storing chemicals (Yeon et al., 2004).

10.4 Safety aspects in electrodeionization

Ecosystems are becoming more and more polluted; hence, sustainable techniques of pollutant removal are needed. The WWT operator is in charge of providing safety during operation in an organized and ongoing manner, including minimizing risk in both typical and atypical operating circumstances. The methods of preparing for business continuity and handling risks, which include the inside as well as the outside of the organization, particularly stakeholder engagement, can be used to carry out this assignment. To both produce and protect wealth, risk control was created. Its use can improve operational effectiveness, assist in the accomplishment of goals, and help the growth of creativity (Tuer and Oulehlová, 2021).

HM, oily substances, herbicides, pesticides, food debris, organic matter, sand, parasitic organisms, dangerous chemicals, antibiotics, and other biological, physical, and chemical pollutants are all present in the water that we use on a daily basis in our homes or in our workplaces as a result of our daily activities. This is referred to as "WW" in most contexts. Before this effluent can be released into rivers or recycled, it must be treated in order to reduce our environmental effect as much as feasible. No matter if they are inside or outdoor, enclosed or open-air, WWT performs this function of treating contaminated water. This WWT is divided into many stages, where employees may be confronted with dangers at any moment, including physical hazards like those associated with slipping or falling, biological hazards like those associated with death by asphyxia poisoned by contaminants, and chemical hazards like those associated with radiation and inhale (Falakh and Setiani, 2018).

The effluent is treated in stations for pumping or transported by tankers before entering the facility. At this point, biological dangers including suffocation, ingestion, and the spread of polluted substances become a concern. There are no chemicals employed in the procedure itself during these pretreatment processes steps. However, when cleaning tools and materials with hydrochloric acid or bleach during service and maintenance activities, there is a substantial danger of exposure to chemicals (Koenig et al., 2008). Risky interaction with these substances is required for the biological therapy to function. Operators run the danger of splashing, which might have fatal repercussions. It is dangerous to come into contact with these substances during cleaning (Chew and Maibach, 2003). The difficulty is to suitably dry or burn the waste material in order to make it stable and hygienic while continually watching out that its drying is not too great, since this increases the danger of creating dust and other problems, such heating itself or explosive reactions (Omer, 2012).

Particularly lately, EDI has shown to be a successful method for removing ionic chemicals from polluted waterways. IEM, MIER, and a DC voltage are used in EDI as part of a continuous ECT for water deionization process. Compared to traditional treatments using acids, bases, and other chemicals, EDI is therefore safer. HPW is produced by EDI, which also enables for effective ion elimination, prevents the development of sludge, and regenerates IER without the use of chemicals. Sustainable EDI does not appear to contain a heating element or movement components; therefore, it requires very little energy, operates extremely cheaply, and is completely safe. Additionally, it does not need IER waste disposal, hazardous material disposal, or chemical trash regeneration (Pan et al., 2017).

10.5 Design aspects in electrodeionization

Operating characteristics including the EP variation, current density, rate of flow, and quality of feed have an impact on how well EDI performs. IER type and shape, IEM, stack arrangement, cell thickness, and chamber length are only a few examples of crucial unit design factors. In order to improve efficiency, many stack setups and building methods have been created. The advancements primarily aim to increase separation efficiency, lower overall opposition, and facilitate the detachment of weakly dissociated elements. According to Hakim et al. (2020), the majority of readily accessible EDI has a mixed-bed arrangement. A research of a newly created configuration, separated-bed with BM, however, revealed that this new design may generate HPW with more efficiency than the mixed-bed. Additionally, by reducing the BM impact and CP at the IEM surface, the elimination efficiency of EDI may be increased. For instance, it was accomplished by utilizing a multilayer BM bed. A further appealing method for power as well as performance optimization is to use a two-stage process or partial EDI with varied voltages and current densities. Heavily and moderately ionized compounds can be effectively removed using this method (Dermentzis, 2010). Fig. 10.1 illustrates a number of design characteristics for the EDI.

The method of design for EDI and traditional ED is quite similar. The primary distinction is that in an EDI system, IER is placed in the DLC and occasionally the CC. For the placement of the CER and AER in the cell, many ideas are applied (Arar et al., 2011). Numerous factors for the layout of the ED device must be chosen based on the shape and substance of the components of the ED stack, such as the IEM, spacer, and

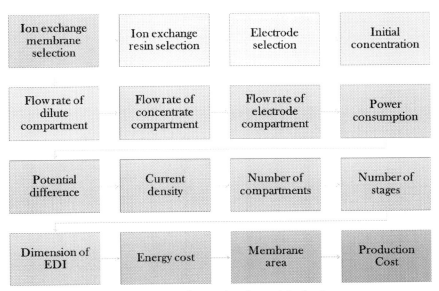

Figure 10.1 Various parameters to design the electrodeionization.

electrodes. In addition to these permanent characteristics, the efficiency and costs of the ED systems are also impacted by variable variables such as input and final water composition, water stacking speed, recuperation ratio, and electrical variables (Ankoliya et al., 2021). A few of the earlier research focused on parameter optimization to increase the effectiveness of the created and fixed-sized ED system. According to research on how voltage applied, surface velocity, and water for feed temperature affect ion elimination, speed has a generally detrimental effect because it shortens ion retention time. Utilizing the normalized sensitivity analysis for different variables, it was discovered that flow speed is the most sensitive variable for the density of current and particular consumption of energy (Valero et al., 2011).

It was believed that when the water that comes in flow rate rises, HPW's resistance rises, implying improved quality. The rise in effective membrane area is responsible for this. However, the rate of rise in water resistance is greater for the rate of flow to rise from 12 to 14 LPH than from 8 to 12 LPH. The rise in the actual IEM area may be attributed to the spike in HPW resistance. This is resulting in higher quality with a rise in intake water flow rate. The quantity of water that is continuously in touch with the IER particle could rise as well due to the greater flow rate. This has the tendency to quicken IER renewal, enabling better ion (salt) transfer from the processed to the CC, from which it was utilized for electrodes rinse to maintain the EP among the two electrodes (Lee et al., 2012).

It was discovered that when the input water level increased, the stack current raised and the resistance of the DLC stream dropped. This happens as a result of the present load and ion load increasing as influent intensity began to rise. According to Lu et al. (2012), a small quantity of hydrogen ions present in the water used for feeding can boost the interstitial solution's electrical conductance and reduce the stack impedance. The feed specifications for EDI are shown in Table 10.1.

An increase in influent temperatures might hasten IE and diffusion and make it easier for ions to move from DLC to CC in the mixture. As a result, increasing the input temperature caused the stack's current to rise and the overall quality of the DLC to greatly improve (Mehenktas and Arar, 2023). Finally, due to its benefits such as continuous operation, high efficacy, stability, and sustainability, EDI may be a feasible and optimistic technique for the treatment of dilute HM wastewater (Peng and Guo, 2020).

Table 10.1 The feed requirements for better performance in electrodeionization.

Variable	Quantity
Conductivity	$< 40\ \mu S/cm$
pH	5–9
Resistivity	16–18 MΩ-cm
Temperature	10°C–35°C
Hardness	<0.25 ppm
Silica	<1 ppm
TOC	<0.5 ppm
Pressure	<60 psi
Chlorine	<0.1 ppm
Sulfide	<0.01 ppm
Iron	<0.01 ppm
Manganese	<0.01 ppm
Carbon dioxide	<10 ppm
Turbidity	0.5–2.0 NTU
Durability	3 years
IEM life	5–10 years

IEM, Ion-exchange membrane.

10.6 Technoeconomic assessment of electrodeionization

Technoeconomic analysis (TEA) is a technique for assessing the technological economic viability. A crucial step toward the industrialization or commercialization of any novel technology or products is TEA at the experimental and development scale. In order to evaluate the practicality of procedures, configurations of systems, bioenergy, and biodegradable products, the assessment may integrate engineering planning, process simulation, financial, and feasibility assessment (Eloka-Eboka et al., 2023). When manufacturing is effectively optimized, TEA can offer financial and energetic benefits while reducing its ecological footprint (Négny and Montastruc, 2022).

TEA is a tool for calculating a prediction of a power plant's efficiency, emission levels, and cost prior to construction. If they depend on current information, take into consideration the expense for labor, materials, and machinery, and base building expenses on existing levels of productivity, these projections can be trusted when utilized for estimating readily accessible systems. Such projections, however, might be more hypothetical and complicated by assumptions when used for technologies that have not yet seen widespread commercialization (Frey and Zhu, 2012). The TEA done takes into account a basic payback and does not take into account the rate of inflation of the corresponding capital expenses for the technology. The main goal of this project's system remodeling strategy is to reduce the related unit electrical expenses since it is conceptualized primarily from the standpoint of improving the efficiency of energy use. The initial expenses, in turn, are in line with the total investment cost of the project (Freire Ordóez et al., 2021).

TEA has been regarded as a crucial instrument for assessing the financial viability of industrial operations. Due to the escalating competitiveness among organizations in multiple sectors, TEA usage has recently been demonstrated to rise exponentially. It is well known that traditional TEA is carried out using software modeling (such as AMIS, Python, Aspen HYSYS, MATLAB, HOMER Pro, Aspen Plus, SysML, FORTRAN, R, and Excel), with no connection or optimization between the procedure and the financial outcome (Chai et al., 2022). Aside from that, manufacturing procedures are being transformed into innovative industries as a result of the coming of the industrial revolution (IR) 4.0. Thus, it is decided that TEA needs to undergo a similar development to innovative industries in order to maintain its honesty. In order to efficiently optimize both procedures and economic factors concurrently, studies have started to include technologies based on data [such as blockchain and artificial intelligence (AI)] into TEA (Goswami et al., 2022). According to the literature evaluations, the most often used data-driven technology in TEA is the genetic algorithm (GA), whereas uses for blockchain technology, machine learning (ML), and artificial neural networks (ANN) are still comparatively rare (Pabuçcu et al., 2023). Other innovative technologies such as cyber-physical systems (CPSs), the Internet of Things (IoTs), cloud computing, big data analysis, digital twins (DTs), and the metaverse have yet to be included into the current TEA. For a precise TEA portrayal of the implementation of intelligent sectors, established expenses related to the technologies mentioned above must also be included (Buchner et al., 2019).

The viability of a full-scale RO/EDI unit with a capacity of 120 m^3/h has been assessed economically in the bench size test. The following presumptions are used in the calculation: The plant runs 330 days per year and has an operational lifespan of 10 years, a price reduction of 10%, a value for salvage of 0, and a price tag of US\$0.66/m^3 for water that has been demineralized, US\$0.038/kWh for power, and US\$0.08/m^3 for water for feeding. The yearly operational expenditures and investment expenses that need to be taken into account are financial concerns. Equipment purchase, plumbing, control system design and setup, installation, training, mobilization and demobilization, management of the project, and birocration are all included in capital costs. The RO package included vessels with pressure, feed pumps, high-pressure pumps, cartridge filtration systems, and RO units. Standard modules, a transfer pumps, and a transformer (a power supply) made comprised the EDI system. A full-scale RO/EDI system has a capital investment of $1,124,816. The overall yearly expenses for running the facility are known as annual operating expenses. The whole of electricity usage equals total energy expenses. The rectifier, RO packaging pump, and EDI packaging pump are some of these electrical consumers. For RO cleaning-in-place, however, the ingredient's cost was the chemicals itself (Jordan et al., 2023; Widiasa and Wenten, 2007). Chemical renewal and the related equipment are not used in EDI technologies, which greatly lowers capital and operational expenses. The total electrical impedance of an EDI module is almost entirely independent of the concentrated water conductance when IER filler is used in the CC and EC of the device. Pumps for salt input or recycling are not required, which reduces the degree of system complexity and possible service time (GRABOWSKI et al., 2006). Since a concentrate recirculating pump may consume almost as much power as EDI's rectifier, this also reduces the running cost. In comparison with the current IE beds, which cost US$0.66/m^3, the particular water price for RO/EDI product water was US$0.53/m^3. The biggest running expenses were power and membrane repair (Wenten and Arfianto, 2013).

10.7 Life cycle analysis of electrodeionization

A crucial instrument for policymakers as well as for business is life cycle analysis (LCA). As a consequence of various business growth and legislative situations, researchers are also keen on projecting forthcoming local and worldwide flows of energy and materials (Pratik et al., 2017). Each product should eventually end up as trash, according to a core concept of LCA. The environmental effects of an item or policy must be considered from "cradle to grave" in order to determine which is more environmentally friendly (Nabavi-Pelesaraei et al., 2022). This covers the potential future (downstream) fate of a product in addition to secondary supplies to the manufacturing procedure, related waste products, and pollutants. Numerical evaluations of material fluxes and changes are made as the evaluation's initial step. Energy fluxes are significant when they include substances such as fuels or combustion byproducts. If done correctly, this activity may be quite beneficial. The information needed to complete this initial stage, nevertheless, is typically not accessible from open sources (Hendrickson et al., 2006; Guinée, 2002). These explanations of theoretical processes might not reflect real practice. Furthermore, "confidential" material is by definition unproven and may contain errors. Such mistakes could go undetected in a lack of a systematic material balance system for accounting (Ayres, 1995; Klöpffer and Grahl, 2014).

LCA was also used to calculate the degree to which the environmental impact of the improved EDI system was reduced. The greenhouse gas (GHG) emissions brought on by the manufacturing, consumption, and disposal of an EDI unit (Pan et al., 2016). The generation of 1 metric tonne (MT) of HPW by EDI is the unit of measurement used to assess such "cradle-to-grave" effects. The dimensions and description of the EDI units were utilized to acquire production and disposal inventory information as well as the running electrical power consumption depending on the optimization of operation variables. The average lifespan of EDI units is anticipated to be 5 years. Based on the global warming potential (GWP) readings of various GHGs published by the Intergovernmental Panel on Climate Change (IPCC) in 2013 (Imasiku et al., 2019), the environmental impact of the optimized EDI system was calculated. The IPCC's 100-year GWP technique, which is frequently employed to quantify the effects of environmental change, normalizes each of the GHGs into the amount of carbon dioxide according to the relative absorption capacity of infrared rays over a one-hundred-year time frame (Yuan et al., 2023).

The need to support the boiler substitute water provides that versatility was made clear by the electricity generation plants' transition to carbon-neutral operations. The most cost-effective configuration for an EDI system that operates at a set flow rate combines big and small units. However, because more resources are needed to manufacture the module, there will be an increase in GHG emissions (Voorspools et al., 2000). The HPW requirement needs to be flexible altered when thermal power stations need to perform for high demand control. In this case, the EDI system's use of both small and big units can lower costs and operational power requirements, and the accompanying GHG emissions from power usage will also go down. When the flexible EDI system is used, operational electrical consumption may be cut by up to 30.21% in comparison with the conventional design, which might result in a yearly reduction of 127,000 MT of emissions of GHGs in China. Therefore, for EDI operation with variable planning, the mixing of small and big modules is crucial and needed. The variables of the model ought to be further altered according to the various EDI unit requirements, and it is also possible to incorporate the CC and ion diffusion (Yuan et al., 2022).

10.8 Conclusion

In this chapter, the effects of EDI on safety and health were examined. The design of EDI was also investigated, along with its technoeconomic assessment and life cycle analysis. Continuous EDI's success depends in large part on its environmental friendliness. Because there does not appear to be a heating element, the effluent associated with resin regeneration is avoided. Additionally, electricity substitutes dangerous chemicals that were previously required for resin regeneration, resulting in lower energy usage, lower operating costs, and complete safety. Additionally, it does not need resin waste disposal or chemical or chemical waste recycling. The major operating costs were for electricity and membrane maintenance. Through such advancements in module design, the cost of EDI modules has significantly decreased over the past few years. But by solving the frame, pipelines, power supply, and control designs — which are yet to be optimized — it is still possible to significantly lower the overall price of EDI devices. Future study must evaluate the running time of the flexible-scheduling EDI system under varied HPW needs in order to calculate a carbon footprint.

References

Alkhadra, M.A., Su, X., Suss, M.E., Tian, H., Guyes, E.N., Shocron, A.N., et al., 2022. Electrochemical methods for water purification, ion separations, and energy conversion. Chemical Reviews 122 (16), 13547–13635. Available from: https://doi.org/10.1021/acs.chemrev.1c00396. Available from: http://pubs.acs.org/journal/chreay.

Alvarado, L., Chen, A., 2014. Electrodeionization: principles, strategies and applications. Electrochimica Acta 132, 583–597.

Alvarado, L., Ramírez, A., Rodríguez-Torres, I., 2009. Cr (VI) removal by continuous electrodeionization: study of its basic technologies. Desalination 249 (1), 423−428.

Alvarado, L., Torres, I.R., Chen, A., 2013. Integration of ion exchange and electrodeionization as a new approach for the continuous treatment of hexavalent chromium wastewater. Separation and Purification Technology 105, 55−62.

Ankoliya, D., Mudgal, A., Sinha, M.K., Davies, P., Licon, E., Alegre, R.R., et al., 2021. Design and optimization of electrodialysis process parameters for brackish water treatment. Journal of Cleaner Production 319. Available from: https://doi.org/10.1016/j.jclepro.2021.128686. Available from: https://www.journals.elsevier.com/journal-of-cleaner-production.

Arar, Ö., Yüksel, Ü., Kabay, N., Yüksel, M., 2011. Removal of Cu^{2+} ions by a microflow electrodeionization (EDI) system. Desalination 277 (1−3), 296−300.

Ayres, R.U., 1995. Life cycle analysis: a critique. Resources, conservation and recycling 14 (3−4), 199−223.

Bilal, M., Rasheed, T., Nabeel, F., Iqbal, H.M.N., Zhao, Y., 2019. Hazardous contaminants in the environment and their laccase-assisted degradation − a review. Journal of Environmental Management 234, 253−264. Available from: https://doi.org/10.1016/j.jenvman.2019.01.001. Available from: http://www.elsevier.com/inca/publications/store/6/2/2/8/7/1/index.htt.

Bouhidel, K.E., Lakehal, A., 2006. Influence of voltage and flow rate on electrodeionization (EDI) process efficiency. Desalination 193 (1−3), 411−421.

Buchner, G.A., Stepputat, K.J., Zimmermann, A.W., Schomäcker, R., 2019. Specifying technology readiness levels for the chemical industry. Industrial and Engineering Chemistry Research 58 (17), 6957−6969. Available from: https://doi.org/10.1021/acs.iecr.8b05693. Available from: http://pubs.acs.org/journal/iecred.

Chai, S.Y.W., Phang, F.J.F., Yeo, L.S., Ngu, L.H., How, B.S., 2022. Future era of techno-economic analysis: insights from review. Front. Sustain 3, 924047.

Chew, A.L., Maibach, H.I., 2003. Occupational issues of irritant contact dermatitis. International Archives of Occupational and Environmental Health 76, 339−346.

Dermentzis, K., 2010. Removal of nickel from electroplating rinse waters using electrostatic shielding electrodialysis/electrodeionization. Journal of Hazardous Materials 173 (1−3), 647−652.

Eloka-Eboka, A.C., Maroa, S., Taiwo, A.E., 2023. Biobutanol from agricultural and municipal solid wastes, techno-economic, and lifecycle analysis. In Advances and Developments in Biobutanol Production. Woodhead Publishing, pp. 171−198.

Falakh, F., Setiani, O., 2018. Hazard identification and risk assessment in water treatment plant considering environmental health and safety practice. In E3S Web of Conferences (Vol. 31, p. 06011). EDP Sciences.

Freire Ordóñez, D., Thorsteinn, H., Ganzer, C., Guillén Gosálbez, G., Mac Dowell, N., Shah, N., 2021. Carbon or Nitrogen-based e-fuels? A comparative techno-economic and full environmental assessment. In 31st European Symposium on Computer Aided Process Engineering (Vol. 50, pp. 1623−1628). Elsevier.

Frey, H.C., Zhu, Y., 2012. Techno-economic analysis of combined cycle systems. Combined Cycle Systems for Near-Zero Emission Power Generation. Woodhead Publishing, pp. 306−328.

Fu, L., Wang, J., Su, Y., 2009. Removal of low concentrations of hardness ions from aqueous solutions using electrodeionization process. Separation and Purification Technology 68 (3), 390−396.

Generous, M.M., Qasem, N.A.A., Akbar, U.A., Zubair, S.M., 2021. Techno-economic assessment of electrodialysis and reverse osmosis desalination plants. Separation and Purification Technology 272. Available from: https://doi.org/10.1016/j.seppur.2021.118875. Available from: http://www.journals.elsevier.com/separation-and-purification-technology/.

Goswami, L., Kayalvizhi, R., Dikshit, P.K., Sherpa, K.C., Roy, S., Kushwaha, A., et al., 2022. A critical review on prospects of bio-refinery products from second and third generation biomasses. Chemical Engineering Journal 448, 137677.

Grabowski, A., Zhang, G., Strathmann, H., Eigenberger, G., 2006. The production of high purity water by continuous electrodeionization with bipolar membranes: Influence of the anion-exchange membrane permselectivity. Journal of Membrane Science 281 (1−2), 297−306. Available from: https://doi.org/10.1016/j.memsci.2006.03.044.

Guinée, J.B. (Ed.), 2002. Handbook on Life Cycle Assessment: Operational Guide to the ISO Standards, Vol. 7. Springer Science & Business Media.

Hakim, A.N., Wenten, I.G., 2014. Advances in electrodeionization technology for ionic separation—a review. Membrane Water Treatment 5 (2), 087.

Hakim, A.N., Khoiruddin, K., Ariono, D., Wenten, I.G., 2020. Ionic separation in electrodeionization system: mass transfer mechanism and factor affecting separation performance. Separation & Purification Reviews 49 (4), 294−316. Available from: https://doi.org/10.1080/15422119.2019.1608562.

Hendrickson, C.T., Lave, L.B., Matthews, H.S., Horvath, A., Joshi, S., McMichael, F. C., et al., 2006. Environmental Life Cycle Assessment of Goods and Services: An Input-Output Approach. Resources for the Future, United States, pp. 1−262. Available from: http://www.taylorandfrancis.com/books/details/9781936331383. Available from: 10.4324/9781936331383.

Imasiku, K., Thomas, V., Ntagwirumugara, E., 2019. Unraveling green information technology systems as a global greenhouse gas emission game-changer. Administrative Sciences 9 (2), 43.

Jeong, C., Ansari, Z., Anwer, A.H., Kim, S.H., Nasar, A., Shoeb, M., et al., 2022. A review on metal-organic frameworks for the removal of hazardous environmental contaminants. Separation and Purification Technology 305, 122416.

Jordan, M.L., Kokoszka, G., Gallage Dona, H.K., Senadheera, D.I., Kumar, R., Lin, Y.J., et al., 2023. Integrated ion-exchange membrane resin wafer assemblies for aromatic organic acid separations using electrodeionization. ACS Sustainable Chemistry and Engineering 11 (3), 945−956. Available from: https://doi.org/10.1021/acssuschemeng.2c05255. Available from: http://pubs.acs.org/journal/ascecg.

Klöpffer, W., Grahl, B., 2014. Life Cycle Assessment (LCA): A Guide to Best Practice. John Wiley & Sons.

Koenig, K.L., Boatright, C.J., Hancock, J.A., Denny, F.J., Teeter, D.S., Kahn, C.A., et al., 2008. Health care facility-based decontamination of victims exposed to chemical, biological, and radiological materials. American Journal of Emergency Medicine 26 (1), 71−80. Available from: https://doi.org/10.1016/j.ajem.2007.07.004.

Kumar, P.S., Varsha, M., Rathi, B.S., Rangasamy, G., 2022. Electrodeionization: fundamentals, methods and applications. Environmental Research. Elsevier.

Lee, J.W., Yeon, K.H., Song, J.H., Moon, S.H., 2007. Characterization of electroregeneration and determination of optimal current density in continuous electrodeionization. Desalination 207 (1−3), 276−285. Available from: https://doi.org/10.1016/j.desal.2006.04.070.

Lee, H.J., Hong, M.K., Moon, S.H., 2012. A feasibility study on water softening by electrodeionization with the periodic polarity change. Desalination 284, 221−227.

Lu, H.X., Bu, S.F., Wang, J.Y., 2012. Effects of feed water conditions on performance of electrodeionization process for removal of Ni^{2+} from dilute solution, Advanced Materials Research, Vol. 502. Trans Tech Publications Ltd, pp. 174−178.

Mehenktaş, C., Arar, Ö., 2023. Application of membrane processes for nitrate (NO^{3-}) removal. Current Chinese Science 3 (1), 42−56.

Muñoz, R., Guieysse, B., 2006. Algal−bacterial processes for the treatment of hazardous contaminants: a review. Water Research 40 (15), 2799−2815.

Nabavi-Pelesaraei, A., Rafiee, S., Mohammadkashi, N., Chau, K.W., Mostashari-Rad, F., 2022. Principle of life cycle assessment and cumulative exergy demand for biodiesel production: farm-to-combustion approach. Synergy Development in Renewables Assisted Multi-Carrier Systems. Springer International Publishing, Cham, pp. 127−169.

Négny, S., Montastruc, L., 2022. Book of Abstracts of the 32nd European Symposium on Computer Aided Process Engineering.

Omer, A.M., 2012. Applications of biogas: state of the art and future prospective. Blue Biotechnology Journal 1 (3), 335.

Pabuçcu, H., Ongan, S., Ongan, A., 2023. Forecasting the movements of Bitcoin prices: an application of machine learning algorithms. arXiv preprint arXiv 2303, 04642.

Pan, S.Y., Lin, Y.J., Snyder, S.W., Ma, H.W., Chiang, P.C., 2016. Assessing the environmental impacts and water consumption of pretreatment and conditioning processes of corn stover hydrolysate liquor in biorefineries. Energy 116, 436−444. Available from: http://www.elsevier.com/inca/publications/store/4/8/3/.

Pan, S.Y., Snyder, S.W., Ma, H.W., Lin, Y.J., Chiang, P.C., 2017. Development of a resin wafer electrodeionization process for impaired water desalination with high energy efficiency and productivity. ACS Sustainable Chemistry and Engineering 5 (4), 2942−2948. Available from: http://pubs.acs.org/journal/ascecg.

Peng, H., Guo, J., 2020. Removal of chromium from wastewater by membrane filtration, chemical precipitation, ion exchange, adsorption electrocoagulation, electrochemical reduction, electrodialysis, electrodeionization, photocatalysis and nanotechnology: a review. Environmental Chemistry Letters 18, 2055−2068.

Pratik, R., Kumar, P., Chaudhary, K., 2017. Life cycle analysis in manufacturing industry—a case study.

Pulkka, S., Martikainen, M., Bhatnagar, A., Sillanpää, M., 2014. Electrochemical methods for the removal of anionic contaminants from water − a review. Separation and Purification Technology 132, 252−271. Available from: https://doi.org/10.1016/j.seppur.2014.05.021. Available from: http://www.journals.elsevier.com/separation-and-purification-technology/.

Qian, F., Lu, J., Gu, D., Li, G., Liu, Y., Rao, P., et al., 2022. Modeling and optimization of electrodeionization process for the energy-saving of ultrapure water production. Journal of Cleaner Production 372. Available from: https://doi.org/10.1016/j.jclepro.2022.133754.

Rathi, B.S., Kumar, P.S., 2020. Electrodeionization theory, mechanism and environmental applications. A review. Environmental Chemistry Letters 18, 1209−1227.

Rathi, B.S., Kumar, P.S., Vo, D.V.N., 2021. Critical review on hazardous pollutants in water environment: occurrence, monitoring, fate, removal technologies and risk assessment. Science of The Total Environment 797, 149134.

Rathi, B.S., Kumar, P.S., Parthiban, R., 2022. A review on recent advances in electrodeionization for various environmental applications. Chemosphere 289, 133223.

Saravanan, A., Yaashikaa, P.R., Senthil Kumar, P., Karishma, S., Thamarai, P., Deivayanai, V.C., et al., 2023. Environmental sustainability of toxic arsenic ions removal from wastewater using electrodeionization. Separation and Purification Technology 317. Available from: https://doi.org/10.1016/j.seppur.2023.123897.

Särkkä, H., Vepsäläinen, M., Sillanpää, M., 2015. Natural organic matter (NOM) removal by electrochemical methods—a review. Journal of electroanalytical chemistry 755, 100−108.

Song, J.H., Yeon, K.H., Moon, S.H., 2007. Effect of current density on ionic transport and water dissociation phenomena in a continuous electrodeionization (CEDI). Journal of Membrane Science 291 (1−2), 165−171.

Su, W., Pan, R., Xiao, Y., Chen, X., 2013. Membrane-free electrodeionization for high purity water production. Desalination 329, 86−92. Available from: https://doi.org/10.1016/j.desal.2013.09.013.

Suman, H., Sangal, V.K., Vashishtha, M., 2021. Treatment of tannery industry effluent by electrochemical methods: a review. Materials Today: Proceedings 47, 1438–1444.

Tušer, I., Oulehlová, A., 2021. Risk assessment and sustainability of wastewater treatment plant operation. Sustainability 13 (9), 5120.

Valentino L., Dunn J.B., Tan E.C., Freeman C.J., Kubic W., Rosenthal A., 2022. Techno-economic analysis and life-cycle assessment of emerging technologies for bioprocessing separations.

Valero, F., Barceló, A., Arbós, R., 2011. Electrodialysis technology: theory and applications. Desalination, Trends and Technologies 28, 3–20.

Verbeek, H.M., Fürst, L., Neumeister, H., 1998. Digital simulation of an electrodeionization process. Computers & Chemical Engineering 22, S913–S916.

Voorspools, K.R., Brouwers, E.A., D D'haeseleer, W., 2000. Energy content and indirect greenhouse gas emissions embedded in 'emission-free' power plants: results for the low countries. Applied Energy 67 (3), 307–330.

Wang, J., Wang, S., Jin, M., 2000. A study of the electrodeionization process—high-purity water production with a RO/EDI system. Desalination 132 (1–3), 349–352.

Wardani, A.K., Hakim, A.N., Khoiruddin, Wenten, I.G., 2017. Combined ultrafiltration-electrodeionization technique for production of high purity water. Water Science and Technology 75 (12), 2891–2899. Available from: https://doi.org/10.2166/wst.2017.173. Available from: http://wst.iwaponline.com/content/75/12/2891.full.pdf.

Wen, R., Deng, S., Zhang, Y., 2005. The removal of silicon and boron from ultra-pure water by electrodeionization. Desalination 181 (1–3), 153–159.

Wenten, I.G., Arfianto, F., 2013. Bench scale electrodeionization for high pressure boiler feed water. Desalination 314, 109–114.

Widiasa, I.N., Wenten, I.G., 2007. Combination of reverse osmosis and electrodeionization for simultaneous sugar recovery and salts removal from sugary wastewater. Reaktor 11 (2), 91–97.

Wood, J., Gifford, J., Arba, J., Shaw, M., 2010. Production of ultrapure water by continuous electrodeionization. Desalination 250 (3), 973–976. Available from: https://doi.org/10.1016/j.desal.2009.09.084.

Wu, G., Kang, H., Zhang, X., Shao, H., Chu, L., Ruan, C., 2010. A critical review on the bio-removal of hazardous heavy metals from contaminated soils: Issues, progress, eco-environmental concerns and opportunities. Journal of Hazardous Materials 174 (1–3), 1–8. Available from: https://doi.org/10.1016/j.jhazmat.2009.09.113.

Yeon, K.H., Song, J.H., Moon, S.H., 2004. A study on stack configuration of continuous electrodeionization for removal of heavy metal ions from the primary coolant of a nuclear power plant. Water Research 38 (7), 1911–1921.

Yuan, Y., Qian, F., Lu, J., Gu, D., Lou, Y., Xue, N., et al., 2022. Design optimization and carbon footprint analysis of an electrodeionization system with flexible load regulation. Sustainability 14 (23), 15957.

Yuan, Y., Lu, J., Gu, D., Lou, Y., Xue, N., Li, G., et al., 2023. Deduction of carbon footprint for membrane desalination processes towards carbon neutrality: a case study on electrodeionization for ultrapure water preparation. Desalination 559. Available from: https://doi.org/10.1016/j.desal.2023.116648. Available from: https://www.journals.elsevier.com/desalination.

Zhang, H., Pap, S., Taggart, M.A., Boyd, K.G., James, N.A., Gibb, S.W., 2020. A review of the potential utilisation of plastic waste as adsorbent for removal of hazardous priority contaminants from aqueous environments. Environmental Pollution 258. Available from: https://doi.org/10.1016/j.envpol.2019.113698. Available from: https://www.journals.elsevier.com/environmental-pollution.

Index

Note: Page numbers followed by "*f*" and "*t*" refer to figures and tables, respectively.

A

Activated carbon (AC), 109–110, 175
Adsorption (AD) method, 15
  developments in, 15, 16*f*
  removal percentage
    of dyes using, 6*t*
    of emerging contaminants using, 11*t*
    of heavy metals using, 8*t*
  wastewater treatment, 110–111
  water softening techniques for, 175
Anion exchange electrodeionization, 90–91
Anion exchange membranes (AEMs), 29–34, 53–54, 64, 69, 79, 104–105, 205–206, 236–237
Anion exchange resin (AER), 53–58, 64, 69, 137–138, 142, 235–236
Anion expulsion, 54–59
  and cation expulsion, 64–68
Arsenate, 54–55
Arsenic (As), 134–135, 138–141
Arsenite, 54–55
Artificial intelligence (AI), 228
Azo dyes, 5

B

Brackish waters, 157, 166
Brine purification methods, 155, 165, 167

C

Cathodic protection chamber, 62–63
Cation exchange electrodeionization, 91–92
Cation exchange membranes (CEMs), 29–34, 53–54, 69, 79–81, 104–105, 236–237
Cation exchange resin (CER), 53–57, 59–62, 62*f*, 69, 71, 137–138, 142, 235–236

Cation expulsion, 59–63
  anion and, 64–68
Chambers, types of, 104–105
Chemical reactions, at electrode, 95–96
Chemisorption, 110–111
Chromium removal, 137–138
Coagulations, 18
Cobalt removal, 142–143
Concentrating compartments (CCs), 57–58, 79, 81–84, 88, 94
Construction, of electrodeionization, 42–43
  chemical reaction happening in, 95–96
  electrode, 88–90
  of electrodialysis, 29–35
  ion-exchange membrane, 37, 85–88
  ion-exchange resin, 81–85
  types of
    anion exchange, 90–91
    cation exchange, 91–92
    mixed bed, 93–95
Continuous EDI (CEDI), 63, 142
Coupling, of electrodeionization with techniques, 226–227
Current–voltage relationship, 44–45

D

Deionized water (DW), 105, 235–236
Desalination (DS), 155, 157
  beneficial method for, 64
  electrodeionization in, 166–168
  techniques for, 158–159, 160*t*
    electrodialysis, 165–166
    membrane distillation, 162–163
    reverse osmosis, 163–165
    solar desalination, 159–162
Design aspects, in EDI, 240–242
Dilute cells (DLCs), 235–236
Dilute chambers, 53–54, 57–58

Index

Dilute compartments (DLCs), 57–58, 62–64, 69–70, 79, 81–84, 88, 94, 104, 123, 137–138

Donnan potential, 45

Dye wastewater treatment approach, 5–6

E

EDI reversal (EDIR), 156–157, 167, 206, 215
 application of, 216t

Electrical resistance, 45–46

Electrochemical methods, in wastewater treatment, 2
 electrocoagulation, 19
 electrodeionization, 20–21
 electrodialysis, 19–20

Electrochemical processes (ECRs), 79

Electrochemical techniques (ECTs), 238

Electrochemistry, 44

Electrocoagulation methods, 19

Electrode, 88–90
 chemical reaction at, 95–96

Electrodeionization (EDI), 38–39
 advantages of, 207–208
 construction of, 42–43
 continuous, 27–28
 deionization performance of, 43
 design aspects in, 240–242
 distinct functioning modes, 39–42
 electrical potential and electroactive materials, 47
 existing technology, limitations of, 208–209
 fundamentals of
  current–voltage relationship, 44–45
  Donnan potential, 45
  electrical resistance, 45–46
  electrochemistry, 44
  limiting current density, 46–47
  transport mechanism, 47
 health aspects in, 237–239
 hybrid technology, 39
 ion cleanup, 39–42
 merits of, 43–44
 methods, 20–21
 operating condition, removal of contaminants using, 40t

past patterns in, 43
 principle of, 39–42
 safety aspects in, 239–240
 technoeconomic assessment of, 243–245

Electrodes, 42, 53–54
 rinse chambers, 57
 rinse containers, 35
 washing chambers, 58–59

Electrodialysis (ED), 28
 brand-new method of, 59–62
 construction of, 29–35, 235
 desalination techniques for, 165–166
 drawbacks of, 35
 vs. electrodeionization with, 119–123, 121t
 and IE technologies, 196
 methods of, 19–20
 operating condition, removal of contaminants using, 30t
 principle of, 28–29
 wastewater treatment, 117
 water softening techniques for, 169–172, 170t

Electrodialysis reversal (EDIR), 105

Electroplating industry (EPI), 107, 144

Electrostatic shielding electrodeionization (ES-EDI), 205–206, 210–215

Expulsion and movement of ions, mechanism for, 54
 anion and cation expulsion, 64–68
 anion expulsion, 54–59
 cation expulsion, 59–63

F

Flocculations, 18

Fluoride-free freshwater, 57–58

Four-chamber EDI technique, 144–150

G

Global warming potential (GWP), 246

Gray water, reusing of, 4

H

Health aspects, in EDI, 237–239

Heavy metals (HMs), 7, 103–104, 129–131
 causes of, 131

effects of, 133–137
ions, 129–130, 144, 145*t*
removal
   arsenic, 138–141
   chromium, 137–138
   cobalt, 142–143
   nickel, 143–144
synthetic sources of, 131–133
Hexavalent compounds, 134
High purity water (HPW), 103, 109, 155–156, 235–236, 240, 242, 246–247
Household potable water demand, 2–3
Hybrid RO/EDI system, 226–227, 227*f*
Hydrogen ions, 237
Hydrometallurgical (HM) industry, 107
Hydroxyl ions, 237

I

Intergovernmental Panel on Climate Change (IPCC), 246
Ion ejection, 54, 64–67
Ion exchange fabrics (IEFs), 195–196
Ion-exchange membranes (IEMs), 27–34, 46, 59–63, 70, 72–73, 79–80, 85–88, 205
Ion exchange (IE) technology, 27, 35–36, 103, 118–119
   construction of, 37
   drawbacks of, 38
   ionic fluxes, 36–37
   principle of, 36–37
   ultrapure water, production of, 195–196
   wastewater treatment, 116
Ion-exchange resin (IER), 27–28, 36, 38–39, 44, 53, 79–85, 130, 205, 223
   adsorption/desorption principles, mass transport, 70–72
   expulsion and movement of ions, 54
      anion and cation expulsion, 64–68
      anion expulsion, 54–59
      cation expulsion, 59–63
   transport mechanisms, 68–70
   transport mechanism in, 47
Ionic separation method, 20–21, 205

L

Life cycle analysis (LCA), 245–247
Limiting current density, 46–47
Liquid flow chambers, 53–54

M

Mass transport, principles of adsorption/desorption affecting, 70–72
Membrane distillation (MD), desalination techniques for, 162–163
Membrane filtration (MF), 116–117
Membrane fouling, 172–173
Membrane-free EDI (MFEDI), 168, 205–206, 218–223
Membrane separation technology, 15–16
Microfiltration (MF), 119
Mining industries (MIs), 106
Mixed bed electrodeionization, 93–95, 137–138, 142
Mixed bed ion-exchange resin (MIER), 81–85, 88, 93–94, 195
Mixed ion-exchange resin (MER), 70
Multiple-effect stack (MES) type evaporator, 162

N

Nanofiltration (NF), 119
   water softening techniques for, 173–174
Nernst-Einstein equation, 47
Nickel removal, 143–144
Nitrate ions, 57
Novel separation opportunities, 193
Nuclear power plants (NPPs), 183–184, 186

O

Organic pollutants, 1, 13

P

Persistent organic pollutants, 13
Pharmaceutical industries (PI), 107
   ultrapure water, production of, 188–189
Polarity reversal (PR), 156–157

Pollutants, in wastewater, 1, 4–5
  contaminants development, 7–13
  dyes, 5–6
  heavy metals, 7
  persistent organic, 13
Pollution-removal technology, 130
Portable water, 1
Pure water generation, 53–54, 186–187

R

Reduction strategy, of water demand, 3–4
Removal, EDI-based HMs
  arsenic, 138–141
  chromium, 137–138
  cobalt, 142–143
  nickel, 143–144
Resin wafer electrodeionization
    (RW-EDI), 167–168, 224–226,
    228
Reverse osmosis (ROs), 111–116
  desalination techniques for, 163–165
  ultrapure water, production of, 194–195
  water softening techniques for, 172–173
Rinse containers, electrode, 35

S

Safety aspects, in EDI, 239–240
Semiconductor industries (SCI), ultrapure
    water, production of, 186–188
Semipermeable membrane, 164
Separation-recovery techniques, 194
Shielding, electrostatic, 210–215
Solar desalination (SDS) systems, 159–162
Synthetic dyes, 5

T

Technoeconomic analysis (TEA), 243–245
Technological industries, 183–184
Transport mechanism, 47, 68–70
Trivalent chromium, 134

U

Ultrafiltration (UF), 119, 237
  polymer-assisted, 226
  ultrapure water, production of, 190–194

Ultrapure water (UPW), 183
  application of, 186
    pharmaceutical industries, 188–189
    power generation, 186–187
    semiconductor industries, 186–188
  manufacturing method, 183–184
  overview of, 184–186
  plant of, 190f
  production of, 189
    electrodeionization, 196–198
    ion-exchange process, 195–196
    reverse osmosis, 194–195
    techniques for, 191t
    ultrafiltration, 190–194
  properties of, 185t
US Environmental Protection Agency and
    the International Agency for
    Research on Cancer, 129

V

VoltaLab, 57

W

Wastewater reuse, 2, 4
Wastewater treatment (WWT), 2,
    103–104, 235
  adsorption, 110–111
  electrochemical methods
    electrocoagulation, 19
    electrodeionization, 20–21
    electrodialysis, 19–20
  electrodialysis, 117
  ion-exchange process, 116
  membrane filtration, 116–117
  in pollutants, 1, 4–5
    contaminants development, 7–13
    dyes, 5–6
    heavy metals, 7
    persistent organic, 13
  reverse osmosis, 111–116
  techniques of, 14–15
    adsorption, 15
    biological methods, 16–18
    coagulation and flocculation, 18
    membrane separation technique,
      15–16

Index　　257

Water breakdown mechanism, 53—54, 141
Water demand, 1—3
　reduction strategy, 3—4
Water-energy nexus, 13—14
Water ionized species, 236
Water mining, 106
Water softening (WS), 168—169
　electrodeionization in, 175—177

techniques for, 169—172, 170t
　adsorption, 175
　electrodialysis, 174—175
　nanofiltration, 173—174
　reverse osmosis, 172—173
Water splitting process, 44—45, 84
Water sustainability, 13—14
White adipose tissue (WAT), 134—135

Printed in the United States
by Baker & Taylor Publisher Services